An Introduction to Mathematical Physiology and Biology

Second Edition

T0296433

An Introduction to Mathematical Physiology and Biology

Second Edition

J. Mazumdar
The University of Adelaide

CAMBRIDGE UNIVERSITY PRESS
Cambridge, New York, Melbourne, Madrid, Cape Town, Singapore, São Paulo

Cambridge University Press
The Edinburgh Building, Cambridge CB2 2RU, UK

Published in the United States of America by Cambridge University Press, New York

www.cambridge.org
Information on this title: www.cambridge.org/9780521641104

First published 1989
Second edition 1999

A catalogue record for this publication is available from the British Library

ISBN-13 978-0-521-64110-4 hardback
ISBN-10 0-521-64110-1 hardback

ISBN-13 978-0-521-64675-8 paperback
ISBN-10 0-521-64675-8 paperback

Transferred to digital printing 2006

This book is dedicated in loving memory to my departed parents and my beloved Guru Bhagavan Sri Sathya Sai Baba.

This book is dedicated in loving memory to my departed parents and my beloved Guru Bhagavan Sri Sathya Sai Baba.

Contents

Preface to the Second Edition

The second and enlarged edition of the book is due to the enthusiastic response to the first edition by teachers and students alike. The present work incorporates many of the suggestions they have put forward.

The general plan, as described in the preface to the first edition, remains unchanged, except for the addition of two new chapters (6 and 7) on *mathematical modelling in epidemiology* and *modelling the AIDS epidemic*. These two chapters have been written by my old friend Professor K.B. Naidu of Sathya Sai Institute of Higher Learning, Bangalore. I am extremely grateful to him for writing these chapters in this book. Although I cannot here express individual thanks to all readers to whom I am indebted for useful critical comments, their communications stimulated me during the past nine years to think of improvements. I hope the new edition will make for easier reading and teaching from the book.

While preparing the second edition of this book, I have received assistance from many people. My heartfelt thanks go out to them all. Especially, I would like to thank TeXAdel, Adelaide for meticulously typesetting the manuscript using LATEX and Anacleto Mernone for his excellent care and cooperation in producing the figures for the book. Finally, I wish to express my appreciation to my wife, Maya, for her personal support and encouragement during the preparation of the revised manuscript.

I also wish to acknowledge the Nanyang Technological University, Singapore for the invitation of a Visiting Professorship in the school of Mechanical and Production Engineering where the final editing of the book is completed.

JAGANNATH MAZUMDAR

ADELAIDE, AUSTRALIA.

Preface to the First Edition

The inspiration for writing this book came from teaching a course on mathematical biology to third year undergraduate students in Applied Mathematics at the University of Adelaide. I started putting together the notes from which this book evolved after nearly a decade of experience in teaching that course. During this period, I have found great difficulty in the choice of textbooks to accompany the lectures. While some books have been written for biologists and medical scientists intending to learn mathematical techniques, there are almost no suitable textbooks which can cater for the needs of a student of an applied mathematics discipline wishing to become aware of the mathematical models used in physiology and biology. This book aims at remedying, at least partially, these defects.

The book contains enough material for a one-semester course on mathematical biology. Our primary users are assumed to be mathematically sophisticated students who have not previously encountered applications of mathematics to physiology and biology; indeed, in most cases will not have studued these subjects at all. It is not primarily intended to cater to the mathematical needs of biological science students although it may be of interest to those life science students who have done mathematics at second year level at university. The text uses mathematical rigour to elucidate physiological and biological phenomena. Much of the necessary mathematics is explained in the text. After completing this book, the student should have a working knowledge of the important mathematical techniques needed to appreciate mathematical modelling in biology and medicine.

A basic problem being studied today by applied mathematicians is how to describe a given biomedical phenomenon in terms of mathematics, i.e., how to construct the mathematical model. I have taken the view that the mathematical model – the statement in mathematical language and symbols – of a problem or phenomenon is central in the application of mathematics in physiology and biology.

The book starts with an introductory chapter on dimensional analysis where only basic concepts and dimensions in empirical equations have been discussed. Chapter 2 discusses the mathematics of diffusion. After explaining the diffusion process in biology, the basic equations are derived and then solutions are obtained using well-known methods for solving boundary value problems. Population dynamics is presented in Chapter 3, where the predator–prey model has been discussed at some length. This chapter also includes the mathematics of autonomous systems of differential equations in the context of the stability of ecosystems. Chapter 4 deals with an introduction to biogeography. A mathematical analysis of wildlife reserves is presented. Chapter 5 presents a discussion of mathematical modelling in pharmacokinetics. The drug distribution among various compartments of the human system has been described with the help of modelling. Chapters 6 and 7 deal with biofluid mechanics which is an important field of study for applied mathematicians. Using the basic principles of mechanics (both solid and fluid) physiological flow characteristics of blood have been extensively discussed. Arterial flow and noninvasive determination of arterial pressure waveforms have been explained. Chapter 8 presents a mathematical modelling approach to problem solving in cardiac mechanics. New approaches to mathematical modelling of the left ventricle are included. Methods of assessing cardiac function in the clinical situation have been discussed. Heart valves, natural and prosthetic, are described in Chapter 9. Also included in this chapter is the spectral analysis of heart sounds using the FFT technique and their physiological significance is explained. Chapter 10 discusses a wide variety of commonly used medical devices.

At the end of each chapter, practice problems have been included which cover the material of that chapter and introduce additional material as well. Some of these problems have been designed to clarify the finer points of the theory and others are straightforward applications of the theory.

Some of the topics presented in the text are old and well-known, but some are relatively new. A word is necessary about the selection of topics in this book and the order in which they have been placed. Although I wanted to include a much wider selection of topics, I found that this was not possible because of the degree of thoroughness that would have been required. I have taken the view that it is more sensible, from the very beginning, to proceed in depth in a few important areas rather than to treat a broader set of topics at a more shallow level. I therefore wish to apologize in advance for this limitation. The exclusion of some topics is partly because of the need to keep the size of the book down to that of a one-semester course book and partly because of my own particular interest, which has dictated the choice of some topics and the exclusion of others.

It is my pleasant duty to express my thanks to many of my colleagues, friends and former students who permitted me to quote their publications and reproduce their figures and data in this book. Among them I wish to mention especially Drs. Dhanjoo Ghista, Trevor Hearn and Paul Stein, Professor Ernie Tuck and Tony Zollo. To Greg Merchant, Des Hill, Greg Noone and Kym Thalassoudis I am thankful for typesetting the manuscript using TeX. I should also thank Professors John Blake and Colin Thompson for going through the manuscript and making a number of valuable comments. Finally, I am grateful to my wife, Maya, for her constant encouragement and sacrifice during the preparation of the manuscript.

In spite of all precautions, errors will no doubt be uncovered while using this book, and I would be most grateful if these errors are brought to my attention so that they can be corrected.

This book is dedicated in loving memory to my father, the late Dr. S.K. Mazumdar, and to my father-in-law, the late Dr. K.K. Datta, both of whom played significant roles during my research career.

J. MAZUMDAR

ADELAIDE, AUSTRALIA.

1

Dimensional Analysis in Mathematical Physiology

1.1. Basic Concepts

The variables which physiologists measure – concentrations, pressures, flow-rates, etc. – are perfectly physical entities whose fundamental nature is in no way altered by their occurrence in a living organism. To deal mathematically with such variables, one must be able to describe them both quantitatively and qualitatively.

Any well-defined physical entity can be described qualitatively by specifying its *dimensions*. Four fundamental dimensions are: mass $[M]$, length $[L]$, time $[T]$, and temperature $[\theta]$. A complete list of various dimensions for physical quantities is given in Table 1.1.

The designation of M, L, T and θ as the fundamental dimensions from which the dimensions of other physical entities are derived is reasonable but by no means unique. It is not even necessary to have four fundamental dimensions. Although the choice of which dimensions to regard as fundamental is, to a

TABLE 1.1. Dimensions of some common physical entities

Physical Entity	Description	Dimensions	SI Units
Mass		M	kg (kilogram)
Length		L	m (metre)
Time		T	s (second)
Temperature		θ	°C
Area	Length Squared	L^2	m^2
Volume	Length Cubed	L^3	m^3
Velocity	Distance per unit Time	LT^{-1}	$m\,s^{-1}$
Acceleration	Rate of Change of Velocity	LT^{-2}	$m\,s^{-2}$
Flow	Volume per unit Time	L^3T^{-1}	$m^3\,s^{-1}$
Density	Mass per unit Volume	ML^{-3}	$kg\,m^{-3}$
Force	Mass × Acceleration	MLT^{-2}	$kg\,m\,s^{-2} = N$
Momentum	Mass × Velocity	MLT^{-1}	$kg\,m\,s^{-1}$
Pressure	Force per unit Area	$ML^{-1}T^{-2}$	$N\,m^{-2} = P$
Work, Energy	Force × Distance	ML^2T^{-2}	$N\,m = J$
Power	Work per unit Time	ML^2T^{-3}	$J\,s^{-1}$
Resistance to Fluid Flow	Pressure Difference per unit Flow	$ML^{-4}T^{-1}$	$kg\,m^{-4}s^{-1}$
Viscosity		$ML^{-1}T^{-1}$	$kg\,m^{-1}s^{-1}$
Fluidity	Inverse of Viscosity	$M^{-1}LT$	$m\,s\,kg^{-1}$
Diffusivity	Coefficient of diffusion	L^2T^{-1}	m^2s^{-1}
Surface Tension	Force per unit Length	MT^{-2}	$kg\,s^{-2}$
Thermal Capacity	Heat per unit Mass-Degree	$L^2T^{-2}\theta^{-1}$	$m^2s^{-2}\,°C^{-1}$
Gas Law Constant	Energy per Mole-Degree	$ML^2T^{-2}\theta^{-1}$	$kg\,m^2s^{-2}\,°C^{-1}$

considerable extent, arbitrary, the M, L, T, θ system is most widely used and will serve our needs perfectly well.

There are some physiological entities which cannot be dealt with mathematically and hence cannot be described dimensionally. Consider, for example, the terms *excitation of a nerve, memory, unconsciousness*, etc. Each of these is important and can be given a reasonably precise definition and yet not one of them can be assigned dimensions.

The importance of knowing the dimensions of each variable is that there are certain rigid rules which specify how dimensional entities can be related to each other. To be valid, any equation which states a general or theoretical relationship between two or more variables must follow these rules for dimensional correctness:

(i) Quantities added or subtracted must have the same dimensions.

(ii) Quantities equal to each other must have the same dimensions.

(iii) Any quantity may be multiplied by or divided by any other quantity without regard to dimensions. However, the resulting product or quotient must have appropriate dimensions so that the above rules are not violated.

(iv) The dimensions of an entity are entirely independent of its magnitude. Hence dx must have the same dimension as x, even though the differential, dx, is infinitesimally small.

(v) Pure numbers, such as Avogadro's number, number of moles, exponents, ratios of two quantities with the same dimensions, e (the base of natural logarithms), have no dimensions.

For example, consider the equation for a *radioactive decay* process where the quantity disappearing at a given time t is proportional to the quantity $Q(t)$, present at that time, i.e.,

$$-\frac{dQ}{dt} = kQ, \tag{1.1}$$

with solution

$$Q = Q_0 e^{-kt}, \tag{1.2}$$

where Q_0 is the amount present at time $t = 0$. In equation (1.1), k is a proportionality constant. Does k have any dimensions?

Assuming that $Q(t)$ is expressed as a mass, and letting $[k]$ stand for *the dimensions of* k, the dimensional equation corresponding to equation (1.1) is

$$[MT^{-1}] = [k][M]$$

which leads to

$$[k] = [T^{-1}]$$

meaning that k must have the dimension of reciprocal time, i.e., k must be a rate. Indeed, this kind of *inverse time* proportionality constant is usually called a *rate constant*.

Consider another example, dealing with the equation for the effect of a drug:

$$E = C\alpha J, \tag{1.3}$$

where E is the effect of a drug, e.g., force of destroying illness with dimensions of force ($[MLT^{-1}]$); C the concentration of drug receptor, e.g., millimoles per litre, with dimensions of $[L^{-3}]$; α the intrinsic activity of the drug, with unknown dimensions, say $[\alpha]$; and J the drug effect per unit concentration, e.g., force per unit concentration of drug receptor, with dimensions $[MLT^{-2}/L^{-3}] = [ML^4T^{-2}]$. Thus, the dimensional relation corresponding to equation (1.3) is

$$[MLT^{-2}] = [L^{-3}]\,[\alpha]\,[ML^4T^{-2}]$$

which implies that α is dimensionless.

1.2. Dimensions in Empirical Equations

An empirical equation is one which has been derived from experimental observations rather than from any underlying theory. It is chosen to represent the trend of the relationship between two or more observed variables, and is therefore the equation of some curve which fits the experimental points.

Suppose we have observed a relationship between a variable x whose dimensions are $[x]$, and a variable y, whose dimensions are $[y]$. We want to find an empirical equation which will describe this relationship. The equation will contain not only x and y but also certain arbitrary constants or parameters a, b, c, ... where numerical values are to be determined from the observed data so that the equation will fit the points.

For example, consider the polynomial

$$y = a + bx + cx^2 + dx^3 + \cdots . \tag{1.4}$$

We see at once that according to our dimensional rules, the dimension of a must be $[y]$, of b must be $[y]/[x]$, of c must be $[y]/[x]^2$, etc., or in general if the nth term is kx^{n-1}, then k must have the dimensions of $[y]/[x]^{n-1}$.

Similarly in the equation for a sum of exponentials

$$y = ae^{k_1 x} + be^{k_2 x} + ce^{k_3 x} + \cdots , \tag{1.5}$$

each k_i must have the dimensions of $1/[x]$ and each of the constants a, b, c, ... must have the dimensions $[y]$.

Occasionally, however, in many biological problems one must use an empirical equation of such a form that the dimensions of the constants cannot be specified in advance of the experiment from the dimensions of x and y. For example

$$y = ax^b; \tag{1.6}$$

now even though the dimensions of x and y are precisely known, the dimensions of x^b will depend upon the numerical value of b, but b is subject to more or less experimental error. In such cases, we must not insist upon the dimensional correctness of the empirical equation.

Consider a simple example. Certain biological variables – odour, basal metabolic rate, etc. – seem to vary directly with the surface area of the body. To facilitate comparison of the metabolic rates of animals of different sizes, it is important to have some way of calculating surface area, for to measure it directly, even in a dead animal, is exceedingly tedious. In contrast, both body length and body weight are relatively easy to measure. Once we know body weight, body volume can also be established from the formula

$$V = \frac{W}{g\rho}, \tag{1.7}$$

where W is the weight of the body in g, ρ is the density in $g\,cm^{-3}$, and g is the acceleration due to gravity. Now let us examine the calculation of surface area in a dog. If all dogs had the same shape, surface area A would be directly proportional to both the square of the body length L, and to the 2/3 power of body volume V, i.e.,

$$A = K_1 L^2 \tag{1.8}$$

and

$$A = K_2 V^{2/3} \tag{1.9}$$

where K_1 and K_2 are proportionality constants. Unfortunately dogs vary considerably in shape so that if the constants are estimated for healthy dogs, equation (1.8) will tend to overestimate the surface area and equation (1.9) will tend to underestimate. The opposite effect will be true for thin dogs.

Hence we will write a dimensionally correct general equation for surface area as a function of body length and body volume in the form

$$A = KL^a \left(V^{1/3}\right)^{2-a} \text{ or } A = KL^a V^{(2-a)/3} \tag{1.10}$$

where a is the exponent of body length and K is a dimensionless constant of proportionality.

As another example, consider the following equation for the rate of transport of a substance, S, which is a substrate for a carrier substance C,

$$V = D(CS_1 - CS_2), \tag{1.11}$$

where V is the transport rate, D the permeability constant, CS_1 the concentration of the carrier substrate complex on side 1 of the membrane and CS_2 is the concentration of the carrier substrate complex on side 2 of the membrane. What are the dimensions of the *permeability constant* and what does it represent?

To answer this question we look at the dimensions of each term appearing in equation (1.11). The transport rate of a substance can be expressed as a quantity of S transported per unit time across a unit area. By letting $[Q]$ be the dimensions of the quantity (however expressed), V would have the dimension

$$[Q][T^{-1}][L^{-2}].$$

The concentrations have the dimensions $[Q][L^{-3}]$, thus

$$[Q][T^{-1}][L^{-2}] = [D][Q][L^{-3}].$$

Therefore

$$[D] = [LT^{-1}],$$

and the *permeability constant* expresses the rate of transfer of a substance across a membrane per unit of membrane area and per unit of concentration difference across the membrane.

1.3. Dimensionless Products

There are some physical entities, such as frequency and angular momentum, whose definitions involve quantities other than mass, length and time. Such entities can be expressed as products involving mass, length and time by algebraic simplification. To each product there is assigned a dimension – that is, an expression of the form $M^a L^b T^c$, where a, b and c are real numbers that may be positive, negative, or zero. When a basic dimension is missing from a product, the corresponding exponent is understood to be zero. When a, b and c are all zero in an expression of the above form, the product is said to be dimensionless.

In general, many dimensionless products may be formed from the variables of a given system. For example, consider the case of a simple pendulum where r denotes the length of the pendulum, m its mass, θ the initial angle of displacement from the vertical, t the period of oscillation, and g the acceleration due to gravity. We wish to find all the dimensionless products among the variables m, r, θ, t and g. Any product of these variables must be of the form

$$m^a \, r^b \, \theta^c \, t^d \, g^e$$

and hence must have dimension
$$[M]^a [L]^b [T]^d [LT^{-2}]^e.$$
Therefore, in order that a product of the above form be dimensionless, we have
$$a = 0,$$
$$b + e = 0,$$
$$d - 2e = 0,$$
which shows there are infinitely many solutions. One dimensionless product is obtained by arbitrarily setting $e = 0$ and $c = 1$, yielding $a = b = d = 0$, and a second independent dimensionless product is obtained when $e = 1$ and $c = 0$, yielding $a = 0$, $b = -1$, and $d = 2$. These give the dimensionless products: $\pi_1 = \theta$ and $\pi_2 = gt^2/r$.

The fundamental result in dimensional analysis that provides for the construction of all dimensionally homogeneous equations from complete sets of dimensionless products is the following theorem.

1.4. Buckingham's Π-Theorem

An equation is dimensionally homogeneous if and only if it can be put into the form
$$F(\pi_1, \pi_2, \dots, \pi_n) = 0 \qquad (1.12)$$
where F is some function of n arguments and $\{\pi_1, \pi_2, \dots, \pi_n\}$ is a complete set of dimensionless products.

Applying *Buckingham's theorem* to the simple pendulum discussed before, it is clear that the two dimensionless products $\pi_1 = \theta$ and $\pi_2 = gt^2/r$ form a complete set in this case. Thus according to the Π-theorem, there exists a function F such that
$$F\left(\theta, \frac{gt^2}{r}\right) = 0. \qquad (1.13)$$
Solving this equation for gt^2/r as a function of θ, it follows that
$$\frac{gt^2}{r} = H(\theta)$$
or
$$t = \sqrt{\frac{r}{g}}\, h(\theta) \qquad (1.14)$$
where h is some function of argument θ.

In fluid mechanics a great many problems may be solved using dimensional analysis. The relevant dimensionless products in fluid mechanical situations may be combined into independent dimensionless groups characterizing the flow. These dimensionless products are precisely defined and named in fluid dynamics.

Buckingham's Π*-theorem* is a method of finding the relevant dimensionless product without knowing the relevant differential equations. However, the method requires that one knows all the relevant variables in any physical problem. For example, for the flow of a viscous incompressible fluid only the equations of motion and continuity are necessary to describe the flow. There are four independent variables u, v, w and p – the three velocity components and the pressure. In addition, there are the fluid properties, density ρ, viscosity μ, and gravitational potential. In fact, the equations of motion in fluid dynamics are obtained from Newton's second law of motion, which states that the product of mass and acceleration of any fluid element is equal to the resultant of all the external body forces acting on the element and to the surface forces acting on the fluid volume due to action of the remaining fluid on the same element. The equations of motion, known as the Navier–Stokes equations, can be written as

$$\rho\left(\frac{\partial u}{\partial t} + u\frac{\partial u}{\partial x} + v\frac{\partial u}{\partial y} + w\frac{\partial u}{\partial z}\right) = -\frac{\partial p}{\partial x} + \mu\nabla^2 u + F_x \qquad (1.15)$$

and similarly for the y and z components. These three equations are written in vector form as

$$\rho\left[\frac{\partial}{\partial t} + (\underset{\sim}{q} \cdot \nabla)\right]\underset{\sim}{q} = -\nabla p + \mu\nabla^2\underset{\sim}{q} + \underset{\sim}{F} \qquad (1.16)$$

where $\underset{\sim}{q}$ is the velocity vector and $\underset{\sim}{F}$ is the gradient of the gravitational potential function.

In the above equation, the terms on the left-hand side represent the inertial forces while the three terms on the right-hand side represent pressure forces, viscous forces and the body forces, respectively. If U is a typical velocity and L is a typical length, the inertial forces are of the order $\rho U^2/L$ and the viscous forces are of the order $\mu U/L^2$. The ratio of these two forces, denoted by Re, is of the order

$$Re = \frac{\rho U L}{\mu} \qquad (1.17)$$

which is a dimensionless quantity, known as the Reynolds number.

Consider now the case of a drag force F that a smooth spherical body experiences in a stream of incompressible fluid. We assume that the five variables F, U, D, ρ, μ are related by a dimensionally homogeneous equation.

This may be expressed as $f(F, U, D, \rho, \mu) = 0$, in which f is an unspecified function. According to *Buckingham's theorem*, f is not actually a function of the five separate variables, but rather a function of a complete set of dimensionless products of the variables. Hence, as explained before, it can be shown that a complete set of dimensionless products of the variables consists of the pressure, $p = F/\rho U^2 D$, and the Reynolds number, $Re = \rho U D/\mu$. Thus, by Buckingham's theorem, the equation can be expressed in the form $f(p, Re) = 0$ or $p = f(Re)$, which indicates that it is possible to plot a curve showing the relationship between p and Re. We will discuss it further in Chapter 8.

Problems

1. Of the following terms, some are not defined mathematically, some are definable but dimensionless, and some have dimensions. Classify them and give dimensions for the dimensional quantities:

   (i) absolute zero,

   (ii) Avogardro's number,

   (iii) cardiac output,

   (iv) "counts per minute" as a measure of radioactive isotope,

   (v) R, the constant in the ideal gas law equation,

   (vi) blood turbulence,

   (vii) white blood cell count,

   (viii) $\log_{10}$ of the injected dose.

2. (i) Show that $A = K L^a V^{(2-a)/3}$, where a and K are dimensionless constants, is a dimensionally correct equation connecting area A, length L and volume V.

   (ii) Suppose you have made a series of measurements of the surface area A, the body length L, and the body volume V of a number of dogs. How could you calculate the most appropriate values of the constants a and K?

   (iii) Use the method to calculate a and K for the following data.

Dog number	1	2	3	4	5	6	7
Length from nose to anus (cm)	74	98	103	51	100	62	76
Body weight (g)	5450	17,250	32,640	3390	25,930	5350	10,150
Surface area (cm²)	3815	8104	10,763	2320	9106	3284	5070

Assume that body weight in grams is equal to body volume in cubic cm.

3. Suppose a sealed liquid container was opened at time $t = 0$ and the liquid was transfered into another container. Let m be the net mass of liquid measured in grams that has been transferred from the container in the time t in minutes. If the relationship between m and t satisfies the equation

$$m = 5 - 5e^{-t},$$

how will this equation be altered if kilograms are used as a mass scale and seconds as a time scale? Show that irrespective of mass and length scales, the measurements will be described by the formula

$$c_1 m = 5 - 5e^{c_2 t}$$

where the constants c_1 and c_2 have the dimensions M^{-1} and T^{-1} respectively.

4. If blood is regarded as a viscous fluid of constant viscosity, using dimensional analysis find the volume of blood flowing per second through a tube of circular cross section (the result is the well-known Poiseuille's formula).

5. Show by dimensional analysis that the period τ of small oscillations of a simple pendulum of length L is given by

$$\tau = c\left(\frac{L}{g}\right)^{1/2}$$

where c is a constant.

6. Use the *Buckingham* Π-*theorem* to obtain the result of the fluid flow through a smooth pipe of circular cross section in the form

$$\frac{\Delta p}{L} = \frac{v^3 \rho^2}{\mu} \phi\left(\frac{v\rho d}{\mu}\right)$$

where $\Delta p/L$ is the pressure drop per unit length of the pipe, and v, ρ, μ and d respectively denote the mean fluid velocity, density, viscosity and diameter of the pipe.

7. Assume the hydrostatic pressure of blood contributes to the total blood pressure in humans. The hydrostatic pressure P is a product of blood

density ρ, height h of blood column between the heart and some lower point in the body, and gravity g. Using dimensional analysis, determine a, b, c such that

$$P = \lambda \rho^a h^b g^c$$

where λ is a dimensionless constant.

8. Assume the force F acting against the fall of a raindrop through air is a product of the viscosity μ of the air, the velocity v, and the diameter r of the drop. Neglecting the density of air, find

$$F = \kappa v^a \mu^b r^c$$

where κ is a dimensionless constant.

9. Using *Buckingham's theorem*, show that in Problem 8 above if one does not neglect the density ρ of the air and if the acceleration due to gravity g is another variable to be considered, then

$$v = \sqrt{rg}\, f\left(\frac{r^{3/2}\, g^{1/2}\, \rho}{\mu}\right)$$

where f is some function of the single-dimensional product.

10. Find the volume flow rate dV/dt of blood flowing in an artery as a function of pressure drop P per unit length of artery, the radius r of the artery, the blood viscosity μ, and the blood density ρ.

2

The Mathematics of Diffusion

2.1. The Process of Diffusion

Consider a solution in which simple molecular diffusion is taking place. A solution consists of a fluid, called the *solvent*, in which some matter, called the *solute*, has dissolved. The composition of the solution is characterized by its mass concentration C, which is the mass of dissolved matter per unit volume of liquid. For example, the water content of cells is approximately 80%, and water is known as the universal biological solvent.

One means by which the solute molecules are dissolved and transported through the solvent is *diffusion*. Thus, diffusion is the process by which matter is transported from one part of a system to another as a result of random molecular motion.

Diffusion processes play a very important role in almost any biological phenomenon. Through diffusion many metabolites are exchanged between a cell and its environment or between the blood stream and tissues.

The diffusion mechanism is the consequence of the thermal motion of the individual solute molecules. The continuous motion of the solvent molecules produces a great many collisions with, say, a given large solute molecule. As a result, pressure fluctuations are produced which in turn impart to the solute molecule a jerky irregular path named a *random walk*. The result of this random walk is a net displacement of the molecule in some direction. The same phenomenon occurs in the case of suspended particles, e.g. pollen in water or emulsions. This random motion is called *Brownian motion* (in honour of the English botanist Robert Brown (1828)). Diffusion is a direct result of the random motion of the molecules in the direction of a gradient. For example in matter, molecules diffuse from areas of high concentration to areas of low concentration while in thermal related areas they diffuse from regions of high temperature to regions of low temperature; i.e., both matter and heat diffuse in the direction of a gradient. Matter diffuses in the direction of a concentration gradient, and heat diffuses in the direction of a temperature gradient.

2.2. Fick's Law of Diffusion

Fick's law of diffusion of matter and also Fourier's law of diffusion of heat were both formulated in the 19th century, before people knew much about atoms and molecules.

Consider a solution in which simple molecular diffusion is occurring, the fluid being otherwise at rest. The mechanism of transport of the solute is governed only by concentration differences. We ask, what is the flux of solute particles going through a unit area in a unit amount of time?

The material flux per unit area is known as the current density and denoted by J. The classical theory of diffusion was founded more than one hundred years ago by the physiologist A. Fick. According to Fick (1855), the *first law of diffusion* is that

$$\text{material flux} = -D \times \text{concentration gradient}$$

or

$$J = -D\frac{dc}{dx} \tag{2.1}$$

where we imagine that the concentration c varies from point to point and depends on the position x only and D is the diffusion coefficient. The diffusion coefficient is a characteristic of the solute of the fluid and in this case is given by

$$D = \frac{KT}{f} \tag{2.2}$$

where K is the Boltzmann constant, T is the absolute temperature and f is a frictional coefficient which depends on the molecular size and shape and the viscosity of the fluid, e.g. for spherical molecules Stokes' law says that $f = 6\pi\mu a$ where a is the radius of the molecule and μ the viscosity of the fluid. The minus sign in equation (2.1) implies that the particle flow proceeds from a high concentration region to a low concentration region.

More generally, the concentration c is a function of position (x, y, z) and time t so that

$$c = c(x, y, z, t). \tag{2.3}$$

For this case Fick's law of diffusion is derived by considering an infinitesimal section of area through a point P. Through this will flow a mass current with components in three space directions. Thus, the current density vector denoted by $\underset{\sim}{J}$ where

$$\underset{\sim}{J} = (J_x, J_y, J_z) = \left(-D\frac{\partial c}{\partial x}, -D\frac{\partial c}{\partial y}, -D\frac{\partial c}{\partial z}\right) \tag{2.4}$$

so that

$$\underset{\sim}{J} = -D\nabla c. \tag{2.5}$$

In principle, D could be a function of (x, y, z) as well as c. When, however, D is a constant, it is called the diffusion constant.

We will now apply the conservation principle to the transportation of solute. The conservation principle is a fundamental principle of nature and all natural phenomena must, whether they are physical, biological or chemical, conform with this principle.

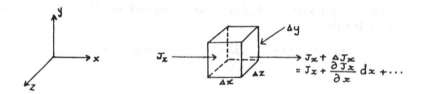

FIGURE 2.1. Conservation of solute transport into and out of a small volume element

Consider a very small cubical volume of solution (as shown in Figure 2.1). Conservation of solute transport into and out of this volume element states that, within a given infinitesimal volume element of a solution in which solute currents exist, whatever the cause of the currents may be, the rate at which matter accumulates or disappears within the region is equal to the net flux across the surface bounding the infinitesimal region. This is stated by the following expression:

$$\left\{ J_x dydz - \left(J_x + \frac{\partial J_x}{\partial x} dx \right) dydz \right\} + \left\{ J_y dxdz - \left(J_y + \frac{\partial J_y}{\partial y} dy \right) dxdz \right\}$$

$$+ \left\{ J_z dxdy - \left(J_z + \frac{\partial J_z}{\partial z} dz \right) dxdy \right\} = \frac{\partial c}{\partial t} dxdydz \qquad (2.6)$$

where the LHS of the above equation is the net diffusive flux of solute across the surface bounding the volume element and the RHS is the increment per unit time of the concentration of the infinitesimal volume element. By simplifying equation (2.6) we obtain

$$\frac{\partial c}{\partial t} + \frac{\partial J_x}{\partial x} + \frac{\partial J_y}{\partial y} + \frac{\partial J_z}{\partial z} = 0 \qquad (2.7)$$

or

$$\frac{\partial c}{\partial t} + \text{div} J = 0, \qquad (2.8)$$

which with the aid of Fick's law (equation (2.4)) reduces to

$$\frac{\partial c}{\partial t} + \frac{\partial}{\partial x} \left(-D \frac{\partial c}{\partial x} \right) + \frac{\partial}{\partial y} \left(-D \frac{\partial c}{\partial y} \right) + \frac{\partial}{\partial z} \left(-D \frac{\partial c}{\partial z} \right) = 0 \qquad (2.9)$$

or

$$\frac{\partial c}{\partial t} + \nabla \cdot (-D\nabla c) = 0. \qquad (2.10)$$

If D is constant then

$$\frac{\partial c}{\partial t} = D\nabla^2 c. \qquad (2.11)$$

For the one-dimensional case we have

$$\frac{\partial c}{\partial t} = D \frac{\partial^2 c}{\partial x^2}, \qquad (2.12)$$

which is called Fick's *second law of diffusion*. Equation (2.11) is usually called the diffusion equation and is similar to the heat equation, where instead of concentration $c(x, t)$ we have temperature $T(x, t)$ and the diffusion constant is replaced by the heat constant.

By applying dimensional analysis to equation (2.12) we find that

$$[D] \rightarrow L^2 T^{-1},$$

because the dimensions of c, x, t are given by

$$[c] \rightarrow ML^{-3},$$
$$[x] \rightarrow L,$$
$$[t] \rightarrow T.$$

The conservation laws can be extended to a finite region. Imagine an arbitrary region of a solution characterized at each point in the region by a concentration c and a current density vector $\underset{\sim}{J}$, each of which varies from point to point in the region. At each point,

$$\frac{\partial c}{\partial t} + \operatorname{div} \underset{\sim}{J} = 0 \qquad (2.13)$$

is satisfied. Assume there are no sources (where matter can be created) or sinks (where matter can disappear) in the region. Let us integrate equation (2.13) over the volume of the region to yield

$$\iiint_V \frac{\partial c}{\partial t} \, dV = - \iiint_V \operatorname{div} \underset{\sim}{J} \, dV = - \oiint_S \underset{\sim}{J} \cdot \underset{\sim}{n} \, dS = - \oiint_S J_n \, dS$$

or

$$\frac{\partial}{\partial t} \iiint_V c \, dV = - \oiint_S J_n \, dS \qquad (2.14)$$

where J_n is the normal component of $\underset{\sim}{J}$, taken to be positive when $\underset{\sim}{n}$ is in the outward normal direction. The quantity $\iiint_V c \, dV$ is just the mass m of solute contained within the volume V at time t. Therefore

$$\frac{\partial m}{\partial t} = -[-J_{\text{in}} + J_{\text{out}}] = J_{\text{in}} - J_{\text{out}} \qquad (2.15)$$

where J_{in} and J_{out} are the rates of mass influx and mass outflux respectively. The above result is called *Fick's principle* and is simply an application of the principle of conservation of mass.

As an example of this principle consider the determination of cardiac output from oxygen consumption by the body. The cardiac output is defined as the steady rate at which blood is ejected by the heart, i.e., the quantity of blood that passes per minute through the circulating system. The cardiac output is an important medical parameter in determining the state of health of the heart.

The cardiac output of the human heart under ordinary conditions of mental and physical rest amounts to 4–5.5 litres per minute, depending upon the size of the individual. This rate, however is greatly increased during excercise. For a young person doing strenous work it may increase to over 20 litres per minute. A number of ingenious methods have been developed, including the ultrasound method, for the measurement of cardiac output.

Let us consider the entire body as one compartment, and let the solute mass be the amount of oxygen in the body. Oxygen is consumed by the body at a steady rate and converted to CO_2. Now, dm/dt represents the steady oxygen consumption rate. Let k denote the steady cardiac output or the mean volume of blood flowing from the heart into the aorta per unit time, which is the blood flow rate into the compartment as well as the blood flow rate leaving the compartment. Let C_a denote the oxygen concentration per volume in the blood of the aorta or arterial side of the circulatory system. Let C_v denote the oxygen concentration per volume in the blood of the vena cava or venous side of the circulatory system. Therefore J_{in} is given by kC_a and J_{out} by kC_v. Hence

$$\frac{dm}{dt} = k(C_a - C_v), \tag{2.16}$$

which upon integration yields

$$m(T) - m(0) = k \int_0^T [C_a(t) - C_v(t)]\, dt. \tag{2.17}$$

The oxygen consumption rate dm/dt as well as C_a and C_v can be determined experimentally, so that the above equation can be used to determine k.

Another example of Fick's principle is the nitrous oxide method for determining cerebral blood flow. In this method an inert gas such as N_2O at constant concentration is inhaled. The concentration of the gas in the arterial blood entering and in the venous blood leaving a given tissue is measured at frequent intervals. Normally the tissue under consideration is the brain.

The blood in one of the veins (left or right jugular vein) is assumed to represent venous outflow from the brain. The typical observations of arterial nitrous oxide concentration C_a and venous nitrous oxide concentration C_v are illustrated in Figure 2.2.

Let m denote the amount of nitrous oxide contained in the cerebral blood at time t, and Q denote the steady rate of blood flow entering or leaving the brain. Then, as before,

$$m(T) - m(0) = Q \int_0^T [C_a(t) - C_v(t)]\, dt. \tag{2.18}$$

The integral on the right represents the total area between the two curves labelled C_a and C_v, and represents the atriovenous nitrous oxide concentration difference between time zero and T.

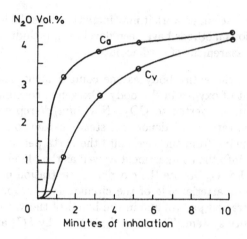

FIGURE 2.2. Typical arterial (C_a) and internal jugular (C_v) curves of N_2O concentration during a ten-minute period of inhalation of 15% N_2O.

When $T \to \infty$, equilibrium is established and C_a and C_v both attain the constant asymptotic value of C_0. This is the concentration of nitrous oxide in cerebral blood at equilibrium. If the nitrous oxide enters the cerebral tissue from the blood by diffusion, C_0 is also the equilibrium concentration in the tissue. Then $m(\infty) = V_B C_0$, where V_B is the volume of the brain. Since $m(0) = 0$, we find that as $T \to \infty$ equation (2.18) becomes

$$V_B C_0 = Q \int_0^\infty [C_a(t) - C_v(t)] \, dt \qquad (2.19)$$

or

$$\frac{Q}{V_B} = C_0 \Big/ \int_0^\infty [C_a(t) - C_v(t)] \, dt. \qquad (2.20)$$

The expression on the right of equation (2.20) is determinable from observations.

It is more usual to express the blood flow to the brain as blood flow per unit mass of the brain. Thus, multiplying the above equations by V_B/M_B, where M_B is the mass of the brain, we get

$$\frac{Q}{M_B} = \frac{C_0 V_B}{M_B} \Big/ \int_0^\infty [C_a(t) - C_v(t)] \, dt$$

$$= M_0 \Big/ \int_0^\infty [C_a(t) - C_v(t)] \, dt \qquad (2.21)$$

where $M_0 = C_0 V_B / M_B$ is the total mass of nitrous oxide taken up by the brain in equilibrium per unit mass of the brain.

2.3. Solutions to the Diffusion Equation in One Dimension

Consider the one-dimensional version of the diffusion equation, namely

$$\frac{\partial c}{\partial t} = D \frac{\partial^2 c}{\partial x^2}, \tag{2.22}$$

where $c = c(x, t)$.

(a) Steady Case

In the steady state case, i.e., $\partial c / \partial t = 0$, equation (2.22) reduces to

$$\frac{\partial^2 c}{\partial x^2} = 0. \tag{2.23}$$

Assuming that c is a function of x only and considering the case of diffusion across a membrane of thickness h, we have that

$$c(x) = Ax + B. \tag{2.24}$$

The boundary conditions are $c(0) = c_1$ and $c(h) = c_2$, therefore the complete solution of equation (2.22) is

$$c = \left(\frac{c_2 - c_1}{h} \right) x + c_1. \tag{2.25}$$

This solution gives a linear concentration profile across the membrane as shown in Figure 2.3.

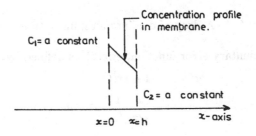

FIGURE 2.3. Concentration profile in membrane

(b) Unsteady Case

Assume equation (2.22) has a solution of the form

$$c(x,t) = \frac{m}{\sqrt{4\pi Dt}}e^{-x^2/4Dt}, \qquad (2.26)$$

where m is a constant. The reason for assuming this form of solution will be evident later on. What is the biological significance of this solution? More mathematically, what are the boundary and initial conditions that it satisfies? To answer these questions consider the case where a *drop* of matter is placed suddenly at the point $x = 0$ at time $t = 0$ in a tube. We shall suppose the tube to be of infinite length in both directions (see Figure 2.4). Very soon after the drop is placed (say a drop of dye into a tube of water), it begins to spread out in both directions. We observe that the total amount of solute diffusing in the medium is given by the expression

$$\int_{-\infty}^{\infty} c(x,t)\, dx = \frac{m}{\sqrt{4\pi Dt}} \int_{-\infty}^{\infty} e^{-x^2/4Dt} dx, \qquad (2.27)$$

where c, the concentration, is defined as the mass of dissolved matter per unit volume of liquid. The integral on the right of the above equation is a form of the error function $\mathrm{erf}(z)$, defined by

$$\mathrm{erf}(z) = \frac{2}{\sqrt{\pi}} \int_0^z e^{-\theta^2}\, d\theta, \qquad (2.28)$$

with the property that

$$\mathrm{erf}(\infty) = 1 \;\Rightarrow\; \int_0^{\infty} e^{-\theta^2}\, d\theta = \frac{\sqrt{\pi}}{2}. \qquad (2.29)$$

$$x = 0$$

FIGURE 2.4. A drop of dye into an infinite tube of water

The complementary error function, $\mathrm{erfc}(z)$, is defined by

$$\mathrm{erfc}(z) = 1 - \mathrm{erf}(z) \qquad (2.30)$$

$$= \frac{2}{\sqrt{\pi}} \int_z^{\infty} e^{-\theta^2}\, d\theta. \qquad (2.31)$$

Now

$$\int_{-\infty}^{\infty} e^{-\theta^2}\, d\theta = 2\int_0^{\infty} e^{-\theta^2}\, d\theta = \frac{2\sqrt{\pi}}{2} = \sqrt{\pi}.$$

Therefore, to solve the integral appearing in the RHS of equation (2.27), we put $\theta = x/\sqrt{4Dt}$ and hence, from above,

$$\int_{-\infty}^{\infty} e^{-x^2/4Dt}\, dx = \sqrt{4Dt\pi}. \tag{2.32}$$

This implies that, from equation (2.27),

$$\int_{-\infty}^{\infty} c(x,t)\, dx = \frac{m}{\sqrt{4\pi Dt}}\sqrt{4\pi Dt} = m. \tag{2.33}$$

Thus m represents the total amount of solute in the system.

From equation (2.26),

$$\frac{c}{m} = \frac{e^{-x^2/4Dt}}{\sqrt{4\pi Dt}}, \tag{2.34}$$

and a graph of c/m vs. x for various values of Dt is shown in Figure 2.5. Since

$$\int_{-\infty}^{\infty} c\, dx = m, \tag{2.35}$$

then

$$\int_{-\infty}^{\infty} \frac{c}{m}\, dx = 1. \tag{2.36}$$

From Figure 2.5 we note that as $t \to 0$, the function gets narrower and more peaked, although the area under it, which is equal to unity, remains the same. In the limit as $t \to 0$, this is reduced to a function which is infinite for $x = 0$ and zero for $x \neq 0$. This function is called the *Dirac delta function* denoted by $\delta(x)$ and defined by

$$\delta(x) = \begin{cases} \infty, & \text{for } x = 0; \\ 0, & \text{for } x \neq 0, \end{cases} \tag{2.37}$$

and has the property that

$$\int_{-\infty}^{\infty} \delta(x)\, dx = 1. \tag{2.38}$$

The Dirac delta function is not a function in the usual sense. In fact, the integral of a function which is zero everywhere except at one point is necessarily zero. It can however be defined in the limiting process as follows (see Figure 2.6):

$$F_\epsilon(x) = \begin{cases} 0, & \text{for } x < -\epsilon; \\ \dfrac{1}{2\epsilon}, & \text{for } -\epsilon \leq x \leq \epsilon; \\ 0, & \text{for } x > \epsilon. \end{cases} \tag{2.39}$$

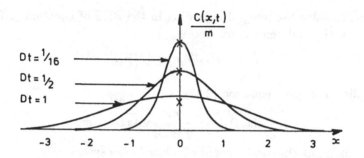

FIGURE 2.5. Variation of $\frac{c}{m}$ with x for various values of Dt

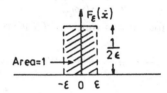

FIGURE 2.6. The Dirac delta function

Then

$$\delta(x) = \lim_{\epsilon \to 0} F_\epsilon(x), \tag{2.40}$$

and also

$$\lim_{\epsilon \to 0} \int_{-\infty}^{\infty} F_\epsilon(x)\,dx = \int_{-\infty}^{\infty} \delta(x)\,dx = 1, \tag{2.41}$$

as the area of the rectangle bounded by $F_\epsilon(x)$ is equal to 1 for all values of ϵ. Since

$$\int_{-\infty}^{\infty} \frac{c}{m}\,dx = \frac{1}{\sqrt{4\pi Dt}} \int_{-\infty}^{\infty} e^{-x^2/4Dt}\,dx, \tag{2.42}$$

it follows that

$$\lim_{t \to 0} \frac{1}{\sqrt{4\pi Dt}} e^{-x^2/4Dt} = \delta(x), \tag{2.43}$$

which leads to the conclusion that (2.26) describes the diffusion of a substance which initially has the distribution

$$c(x,0) = m\delta(x). \tag{2.44}$$

This represents the diffusion of a thin slab of solute placed at $x = 0$ in an infinite medium. Because the solute mass is finite and the medium is of infinite extent the concentration must satisfy the boundary conditions

$$c(\pm\infty, t) = 0, \tag{2.45}$$

as otherwise the solute mass would be infinite. We see that the solution

$$c(x, t) = \frac{m}{\sqrt{4\pi Dt}} e^{-x^2/4Dt}$$

has this property.

This solution with $m = 1$ is called the *unit one-dimensional source solution*. When the solute mass is placed at $x = x_0$, the unit one-dimensional source solution is

$$c(x, t) = \frac{m}{\sqrt{4\pi Dt}} e^{-(x-x_0)^2/4Dt} \tag{2.46}$$

where $m = 1$.

2.4. Diffusion Through a Membrane

We shall now consider in a more precise way the diffusion of a solute through a plane membrane of thickness h, the two sides of which are maintained at constant concentrations c_1 and c_2 respectively (see Figure 2.7). Assume initially that there is no solute present in the membrane. Thus mathematically $c(x, t)$ satisfies

$$\frac{\partial c}{\partial t} = D \frac{\partial^2 c}{\partial x^2} \tag{2.47}$$

in the region $0 < x < h$ and is subject to the boundary conditions

$$c(0, t) = c_1, \qquad t > 0 \tag{2.48a}$$
$$c(h, t) = c_2, \qquad t > 0 \tag{2.48b}$$

and initial condition

$$c(x, 0) = 0. \tag{2.48c}$$

(a) Solution by the Method of Separation of Variables

We now use the method of separation of variables to solve equation (2.47) subject to the above boundary and initial conditions. Assume that

$$c(x, t) = X(x)T(t). \tag{2.49}$$

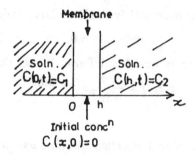

FIGURE 2.7. Diffusion across a membrane of thickness h

Then by substituting this into equation (2.47) and rearranging we get

$$\frac{1}{DT}\frac{dT}{dt} = \frac{1}{X}\frac{d^2X}{dx^2} = -\lambda^2 \quad \text{(say)}. \tag{2.50}$$

The constant $-\lambda^2$ must be negative because if it were positive, say μ, then the solution would be

$$T = ae^{\mu t}$$

$$X = c_1 e^{\sqrt{\mu/D}x} + c_2 e^{-\sqrt{\mu/D}x},$$

which implies that as t increases $c(x,t)$ is unbounded. From equation (2.50) we get that

$$\frac{dT}{dt} = -\lambda^2 DT, \tag{2.51a}$$

$$\frac{d^2X}{dx^2} = -\lambda^2 X, \tag{2.51b}$$

which gives

$$X = \begin{cases} A_1 x + A_2, & \text{for } \lambda = 0; \\ B_1 \sin \lambda x + B_2 \cos \lambda x, & \text{for } \lambda \neq 0, \end{cases} \tag{2.52}$$

and

$$T = \begin{cases} \text{const}, & \text{for } \lambda = 0; \\ B_3 e^{-\lambda^2 Dt}, & \text{for } \lambda \neq 0. \end{cases} \tag{2.53}$$

Therefore

$$c(x,t) = \begin{cases} a_1 x + a_2, & \text{for } \lambda = 0; \\ (A \sin \lambda x + B \cos \lambda x) e^{-\lambda^2 Dt}, & \text{for } \lambda \neq 0. \end{cases} \tag{2.54}$$

By applying the boundary conditions (2.48a) and (2.48b) to the solution for $\lambda = 0$, and using the principle of superposition of solutions, we write

$$c(x,t) = c_1 + \frac{(c_2 - c_1)}{h}x + (A \sin \lambda x + B \cos \lambda x)\, e^{-\lambda^2 Dt}. \qquad (2.55)$$

Applying the boundary conditions to the above expression implies that

$$B = 0$$

and

$$\lambda = \lambda_n = \frac{n\pi}{h}, \qquad n = 1, 2, 3, \dots,$$

therefore

$$c(x,t) = c_1 + \frac{(c_2 - c_1)}{h}x + \sum_{n=1}^{\infty} A_n \sin \lambda_n x e^{-\lambda_n^2 Dt} \qquad (2.56)$$

where

$$\lambda_n = \frac{n\pi}{h}.$$

The initial condition, (2.48c), requires that

$$-c_1 - \frac{(c_2 - c_1)}{h}x = \sum_{n=1}^{\infty} A_n \sin \lambda_n x. \qquad (2.57)$$

Using the Fourier series expansion yields

$$\left.\begin{aligned}
A_n &= \frac{2}{h}\int_0^h \left[-c_1 - \frac{(c_2 - c_1)}{h}x\right] \sin \frac{n\pi x}{h}\, dx \\
&= \frac{2}{n\pi}\left[(-1)^n c_2 - c_1\right].
\end{aligned}\right\} \quad n = 1, 2, 3, \dots. \qquad (2.58)$$

Thus the final solution is

$$c(x,t) = c_1 + \frac{(c_2 - c_1)}{h}x + \frac{2}{\pi}\sum_{n=1}^{\infty} \frac{1}{n}\left((-1)^n c_2 - c_1\right) \sin \frac{n\pi x}{h} e^{-n^2\pi^2 Dt/h^2}. $$
$$(2.59)$$

From this expression, we can readily calculate $D(\partial c/\partial x)|_{x=0}$ which is the current density J_0 or rate at which the diffusing substance emerges at the interface $x = 0$ per unit area per unit time. We find that

$$J_0 = \frac{D}{h}(c_2 - c_1) + \frac{2D}{h}\sum_{n=1}^{\infty}\left((-1)^n c_2 - c_1\right) e^{-n^2\pi^2 Dt/h^2}. \qquad (2.60)$$

The first term of equation (2.60) is the steady state flux while the series term is the transient flux which is significant only for short times. Equation (2.59) is a useful representation of the solution except for very short times when a great many terms contribute to it. However, because the exponential coefficient appearing in equations (2.59) and (2.60) is proportional to n^2 the terms in the series with large n decay very rapidly with time. Thus as a good approximation we may retain only the first terms. Therefore for $n = 1$

$$c(x,t) \approx c_1 + \frac{(c_2 - c_1)}{h}x - \frac{2}{\pi}(c_1 + c_2) \sin \frac{\pi x}{h} e^{-\pi^2 Dt/h^2} \qquad (2.61)$$

and

$$J_0 \approx \frac{D}{h}(c_2 - c_1) - \frac{2D}{h}(c_1 + c_2)e^{-\pi^2 Dt/h^2} \qquad (2.62)$$

The dimensions of the variables are as follows:

$$[D] \rightarrow L^2 T^{-1},$$
$$[c] \rightarrow ML^{-3},$$
$$[x] \rightarrow L,$$
$$[t] \rightarrow T,$$

which implies that h^2/D has the dimension of time; this is the characteristic time of the membrane.

We will now turn to the biological application. Consider that the slab in question represents a cell in a large volume of solute with fixed concentration c_0. In other words, the region $0 \leq x \leq h$ represents the cell, while the regions $x < 0$ and $x > h$ represent the cell exterior. The slab thickness h can be thought of as a characteristic length or size of the cell.

Since the solute has a fixed concentration c_0, then $c_1 = c_2 = c_0$ so we have from equation (2.59)

$$c(x,t) = c_0 \left(1 - \frac{4}{\pi} \sum_{n=1,3,5,...} \frac{1}{n} \sin \frac{n\pi x}{h} e^{-n^2 \pi^2 Dt/h^2} \right). \qquad (2.63)$$

We can now deduce the fractional equilibration ratio of solute mass (i.e., diffusing substance) in the cell $m(t)$ to the solute mass in the cell at equilibrium $m_0 = m(\infty)$. Hence

$$\frac{m}{m_0} = \int_0^h c(x,t)\, dx \bigg/ \int_0^h c(x,\infty)\, dx$$

$$= 1 - \frac{4}{\pi h} \sum_{n=1,3,5,...} \frac{1}{n} e^{-n^2 \pi^2 Dt/h^2} \int_0^h \sin \frac{n\pi x}{h}\, dx$$

$$= 1 - \frac{8}{\pi^2} \sum_{n=1,3,5,...} \frac{1}{n^2} e^{-n^2 \pi^2 Dt/h^2}. \qquad (2.64)$$

Thus except for very short intervals

$$\frac{m}{m_0} \approx 1 - \frac{8}{\pi^2}e^{-\pi^2 Dt/h^2}. \tag{2.65}$$

(b) Solution by the Laplace Transformation Method (for small values of the time)

The method used in the preceding subsection may be said to be the immediate consequences of Fourier's classical work. Another method, known as the Laplace transformation method, has also been well developed and is especially adapted to the solution of diffusion problems for small values of the time.

Suppose that we have to solve the diffusion equation (2.47) with initial condition (2.48c). We apply the Laplace transformation with respect to t to (2.47), that is, multiply by e^{-pt} and integrate with respect to t from 0 to ∞. This gives

$$\int_0^\infty e^{-pt}\frac{\partial^2 c}{\partial x^2}\,dt - \frac{1}{D}\int_0^\infty e^{-pt}\frac{\partial c}{\partial t}\,dt = 0 \tag{2.66}$$

which after integration yields

$$\frac{d^2\bar{c}}{dx^2} - \frac{p}{D}\bar{c} = 0, \quad 0 < x < h \tag{2.67}$$

where $\bar{c}$ denotes the Laplace transform of the function $c(x, t)$ and is a function of x and p, p being a number whose real part is positive and large enough to make the integrals in (2.66) convergent.

The Laplace transformation thus reduces the partial differential equation to an ordinary differential equation. The equation for $\bar{c}$ derived in this way can be solved with the help of two boundary conditions.

Consider the case when there is no concentration gradient at $x = 0$, and $x = h$ is maintained at constant concentration c_0. These two boundary conditions after application of Laplace transformation reduce to

$$\frac{d\bar{c}}{dx} = 0, \quad x = 0, \quad t > 0 \tag{2.68a}$$

$$\bar{c} = \frac{c_0}{p}, \quad x = h, \quad t > 0. \tag{2.68b}$$

The solution of (2.67) subject to (2.68) is

$$\bar{c}(x, p) = \frac{c_0}{p}\frac{\cosh(qx)}{\cosh(qh)} \tag{2.69}$$

where we have written q^2 for p/D. Taking the Laplace inversion in (2.69) yields the usual result in the form discussed in the previous section. However, we will now use a device which gives an alternative form of solution which is often more useful than those earlier, especially for small values of the time.

Expressing the hyperbolic functions in equation (2.69) in terms of negative exponentials, and expanding in a series by the binomial theorem, we obtain

$$\bar{c} = c_0 \frac{e^{qx} + e^{-qx}}{p\, e^{qh}(1 + e^{-2qh})}$$

$$= \frac{c_0}{p}\left(e^{-q(h-x)} + e^{-q(h+x)}\right) \sum_{n=0}^{\infty} (-1)^n e^{-2nqh}$$

$$= \frac{c_0}{p} \sum_{n=0}^{\infty} (-1)^n e^{-q[(2n+1)h-x]} + \frac{c_0}{p} \sum_{n=0}^{\infty} (-1)^n e^{-q[(2n+1)h+x]}. \qquad (2.70)$$

Using the fact that the Laplace inversion is given by

$$\mathcal{L}^{-1}\left\{\frac{e^{-qx}}{p}\right\} = \mathrm{erfc}\left(\frac{x}{2\sqrt{Dt}}\right) \qquad (2.71)$$

we finally obtain

$$c(x,t) = c_0 \sum_{n=0}^{\infty} (-1)^n \,\mathrm{erfc}\left(\frac{(2n+1)h - x}{2\sqrt{Dt}}\right)$$

$$+ c_0 \sum_{n=0}^{\infty} (-1)^n \,\mathrm{erfc}\left(\frac{(2n+1)h + x}{2\sqrt{Dt}}\right). \qquad (2.72)$$

This series converges quite rapidly except for large values of Dt/h^2. It is thus complementary to the solution of the previous subsection, which is most rapidly convergent for large values of the time. For example, if $Dt/h^2 = 1$ we have from (2.72) for $x = 0$

$$c = 2c_0 \left[\mathrm{erfc}\left(\frac{1}{2}\right) - \mathrm{erfc}\left(\frac{3}{2}\right) + \mathrm{erfc}\left(\frac{5}{2}\right) - \cdots\right]$$

$$= 0.89\, c_0. \qquad (2.73)$$

(c) Solution by Goodman's Integral Method

An approximate analytic solution obtained by Goodman (1958) satisfies not the partial differential equation itself but its integrated form with respect to the space variable. The method was originally studied in the context of heat conduction problems, and is called the heat-balance integral method, by which good approximate solutions to nonlinear transient heat conduction problems

can be obtained. Since the diffusion problems are analogous to heat conduction problems, the method is equally applicable to diffusion problems. The
method reduces the nonlinear boundary value problem to an ordinary initial
value problem whose solution can frequently be expressed in closed analytical
form.

Let us explain the method to obtain the solution to a simple heat conduction problem. Assume there is a semi-infinte slab extending over positive
x. Initially, the temperature T is T_∞, and at the surface $x = 0$ the heat flux
$F(t)$ is given for time $t > 0$. If α is the thermal diffusivity, the heat conduction
equation is

$$\frac{\partial T}{\partial t} = \alpha \frac{\partial^2 T}{\partial x^2}, \quad x > 0, \quad t > 0 \tag{2.74}$$

which is analogous to equation (2.12).

The heat flux boundary condition is given by

$$\kappa \frac{\partial T}{\partial x} = -F(t), \quad x = 0, \quad t > 0 \tag{2.75}$$

where κ is the thermal conductivity. A quantity $h(t)$ is now defined to be the
thermal layer thickness. Its properties are such that for $x > h(t)$, the slab, for
all practical purposes, is at an equilibrium temperature and there is no heat
transferred beyond this point. It is analogous to boundary layer thickness in
fluid mechanics. If equation (2.74) is multiplied by dx and integrated from 0 to
h, the resulting equation is called the heat-balance integral. The original heat
conduction equation will, therefore, be satisfied only on the average.

The heat-balance integral thus obtained is

$$\frac{d}{dt}(\theta - hT_\infty) = \alpha \left[\frac{\partial T}{\partial x}(h, t) - \frac{\partial T}{\partial x}(0, t) \right] \tag{2.76}$$

where

$$\theta = \int_0^{h(t)} T \, dx. \tag{2.77}$$

But since there is no heat transferred beyond $x = h$,

$$\frac{\partial T}{\partial x}(h, t) = 0. \tag{2.78}$$

Let us assume that T can be represented by a second-degree polynomial of the
form

$$T = a + bx + cx^2 \tag{2.79}$$

where the coefficients may depend on t. Using equations (2.75), (2.78) and the condition that $T(x,t) = T_\infty$ at $x = h$, the temperature profile ultimately takes the form

$$T = T_\infty + \frac{F}{2\kappa h}(h - x)^2 \qquad (2.80)$$

which, when substituted in (2.77), yields

$$\theta = hT_\infty + \frac{h^2 F}{6\kappa}. \qquad (2.81)$$

With the help of equations (2.75), (2.78) and (2.81), the heat-balance integral equation (2.76) finally reduces to the following ordinary differential equation in h:

$$\frac{d}{dt}(h^2 F) = 6\alpha F. \qquad (2.82)$$

Using the initial condition $h(0) = 0$, the above differential equation yields the solution

$$h = \sqrt{6\alpha}\left[\frac{1}{F(t)}\int_0^t F(\tau)\,d\tau\right]^{\frac{1}{2}}. \qquad (2.83)$$

If $F(t)$ is constant, this reduces to

$$h = \sqrt{6\alpha t}. \qquad (2.84)$$

The surface temperature is thus given by

$$T(0,t) = T_\infty + \frac{1}{\kappa}\sqrt{\frac{3\alpha}{2}}\left[F(t)\int_0^t F(\tau)\,d\tau\right]^{\frac{1}{2}}. \qquad (2.85)$$

If $F(t)$ is constant, this becomes

$$T(0,t) = T_\infty + \frac{1}{\kappa}\sqrt{\frac{3\alpha t}{2}}\,F. \qquad (2.86)$$

The exact solution of this problem is given by Carslaw and Jaeger (1959), and for constant F reduces to

$$T(0,t) = T_\infty + \frac{1}{\kappa}\sqrt{\frac{4\alpha t}{\pi}}\,F. \qquad (2.87)$$

By comparing equations (2.86) and (2.87) it is seen that the results are of the same form and differ only by a numerical factor. The error may be minimized by considering a higher-degree polynomial in (2.79).

It is thus seen that the Goodman integral method is a relatively simple method to use though it is not always clear how accuracy can be improved.

2.5. Convective Transport

When a solute is in a moving liquid, it is transported by the flow. The resulting motion of the solute is called convective transport. This transport is additional to the diffusive motion that the solute undergoes.

We have an additional current density called *convective current density* given by

$$J_{\text{conv}} = c\underset{\sim}{q} \qquad (2.88)$$

where c is the concentration of the solute and $\underset{\sim}{q}$ is the flow velocity vector (u, v, w), in addition to the diffusional current density,

$$J_{\text{diff}} = -D\nabla c. \qquad (2.89)$$

Combining these two current densities we have

$$J_{\text{total}} = c\underset{\sim}{q} - D\nabla c. \qquad (2.90)$$

Now Fick's law,

$$\frac{\partial c}{\partial t} + \text{div } \underset{\sim}{J} = 0, \qquad (2.91)$$

reduces to

$$\frac{\partial c}{\partial t} + \text{div}(c\underset{\sim}{q}) - \text{div}(D\nabla c) = 0 \qquad (2.92a)$$

or

$$\frac{\partial c}{\partial t} + \frac{\partial}{\partial x}(cu) + \frac{\partial}{\partial y}(cv) + \frac{\partial}{\partial z}(cw) = D\nabla^2 c. \qquad (2.92b)$$

The velocity vector $\underset{\sim}{q}$ must either be known and given or this equation must be supplemented by the fluid flow equation for $\underset{\sim}{q}$. When the velocity vector $\underset{\sim}{q}$ is constant we have

$$\frac{\partial c}{\partial t} + (\underset{\sim}{q} \cdot \nabla) c = D\nabla^2 c. \qquad (2.93)$$

This equation is also true for the more general case when the liquid is incompressible, i.e., when

$$\nabla \cdot \underset{\sim}{q} = 0.$$

In one dimension, we have the one-dimensional convective diffusive equation, namely

$$\frac{\partial c}{\partial t} + u\frac{\partial c}{\partial x} = D\frac{\partial^2 c}{\partial x^2}. \qquad (2.94)$$

By introducing a change of variables given by

$$\xi = x - ut,$$
$$\eta = t$$

we get

$$\frac{\partial c}{\partial \eta} = D \frac{\partial^2 c}{\partial \xi^2}. \tag{2.95}$$

Thus the solution of the convective diffusion equation can be written as $c(x - ut, t)$, where $c(x, t)$ is the solution of the one-dimensional diffusion equation. The application of such a solution could be to study the flow of ions inside an *axon*, a fibre of a nerve cell.

2.6. Sedimentation

The masses of particles or large molecules, such as proteins, are usually determined by observing their motion under the influence of outside forces such as gravity or centrifugal force.

In a gravitational field, the terminal velocity of a molecule is determined by the balance between the force of gravitation and the frictional resistance of the fluid medium to the motion of the molecule. The force of gravitation, which is the difference between the weight of the molecule and the buoyant force due to the displacement of the fluid medium, is

$$mg - V\rho_0 g \tag{2.96}$$

where m is the mass of the molecule, V is the volume ($= m/\rho$) and ρ_0 is the density of the fluid medium. Expression (2.96) can also be written as

$$mg \left(1 - \frac{\rho_0}{\rho}\right). \tag{2.97}$$

When this force balances with the frictional force fu, the uniform velocity u at which the molecule *sediments* will be

$$u = \frac{mg}{f} \left(1 - \frac{\rho_0}{\rho}\right). \tag{2.98}$$

2.7. Ultracentrifugation

The ultracentrifuge is an instrument for spinning biochemical solutions at very large angular velocities, thereby subjecting any solute or suspension in the solution to a strong centrifugal force field. With it, the constituents of the cell are dispersed.

In this case g in the above equation has to be replaced by $\omega^2 r$, in which ω is the angular velocity of the centrifuge and r is the distance between the molecule and the centre of rotation.

The sedimentation constant

$$s = \frac{u}{\omega^2 r} \tag{2.99}$$

is a characteristic constant for a given molecular species in a given solvent.

The customary unit of s is 1 svedberg $\equiv 10^{-13}$ s. Thus we have from equations (2.98) and (2.99),

$$s = \frac{m}{f}\left(1 - \frac{\rho_0}{\rho}\right). \tag{2.100}$$

In this case of sedimentation, we have the total solute current density $\underset{\sim}{J}$ given by

$$\underset{\sim}{J} = -D\nabla c + c\underset{\sim}{q} \tag{2.101}$$

where the $c\underset{\sim}{q}$ term is the convective current or sedimentation current density.

The equation of continuity, taking into account the effect of the centrifugal field as well as diffusion or solute transport, reduces to

$$\frac{\partial c}{\partial t} + \operatorname{div} \underset{\sim}{J} = 0 \tag{2.102}$$

which implies that

$$\frac{\partial c}{\partial t} = \frac{1}{r}\frac{\partial}{\partial r}\left(Dr\frac{\partial c}{\partial r} - s\omega^2 r^2 c\right) \tag{2.103}$$

where $c = c(r,t)$ and $\underset{\sim}{q} = s\omega^2 r\hat{r}$. The above equation is known as the *Lamm equation*.

The determination of the sedimentation coefficient s is an important task for biologists. This determination depends in turn on observing the velocity of a solute molecule in a centrifugal force field. This is done by optical methods.

Because D and s depend in general on c, the Lamm equation cannot be solved explicitly. However, with the help of certain simplifying assumptions, this equation can be solved.

Problems

1. Show that

$$c(x,t) = \frac{c_0}{2}\left(1 - \text{erf}\left\{\frac{x}{\sqrt{4Dt}}\right\}\right)$$

satisfies the differential equation of diffusion in one dimension, with boundary condition

$$c(0,t) = c_0/2,$$

and initial condition

$$c(x,0) = 0 \qquad \text{for } x > 0.$$

You are given

$$\text{erf}(z) = \frac{2}{\sqrt{\pi}}\int_0^z e^{-\eta^2}\, d\eta.$$

If $D = 10^{-3}\,\text{cm}^2\,\text{s}^{-1}$, find at what position the concentration will reach a level of 1% of c_0 after 10 seconds. The following values are given:

z	1.386	1.452	1.535	1.645	1.821
$\text{erf}(z)$	0.95	0.96	0.97	0.98	0.99
$(2/\sqrt{\pi})\exp(-z^2)$	0.1653	0.1369	0.1071	0.0754	0.0409

2. (i) Verify that

$$c(x,t) = \frac{m}{\sqrt{4\pi Dt}}e^{-x^2/4Dt}, \qquad -\infty < x < \infty, \quad t > 0,$$

is a solution of the one-dimensional diffusion equation. State the functional form of $c(x,t)$ when $t = 0$.

(ii) Suppose at $x = 2$, $0.1\,\text{kg}$ of solute is injected into the one-dimensional system $0 \le x < \infty$ which has a nonpermeable wall at $x = 0$. What boundary condition applies at $x = 0$? What is the concentration $c(x,t)$ at any position $x > 0$ and time t?

3. The concentration $c(x,t)$ of solute in a solution depends on one space coordinate, and on time. Deduce the corresponding expression for the flux, if the solution (solute and solvent) has a uniform velocity U in the direction x. Hence show that the differential equation for $c(x,t)$ in this situation of convection–diffusion is

$$\frac{\partial c}{\partial t} + U\frac{\partial c}{\partial x} = D\frac{\partial^2 c}{\partial x^2}.$$

Verify that a time-dependent solution for the above equation is

$$c(x,t) = \frac{c_0}{2} \left(1 - \text{erf} \left\{ \frac{x - Ut}{\sqrt{4Dt}} \right\} \right),$$

with the initial condition

$$c(x,0) = \begin{cases} 0 & \text{for } x > 0; \\ c_0 & \text{for } x < 0. \end{cases}$$

4. A mica sheet of thickness L with a number of "pores" (cylindrical holes of length L, radius a) separates a reservoir of water, containing concentration c_1 of tritium oxide, from another reservoir of water with no tritium oxide. The pressure difference between the reservoirs causes a flow of water with mean velocity U in the tube. Write down the convection–diffusion equation for $c(x,t)$ of tritium oxide for the given end condition. Show that the flux (per unit area) of tritium oxide in the tube is

$$J = \left(\frac{D}{L} \right) \frac{\alpha c_1 e^\alpha}{e^\alpha - 1} \quad \text{where} \quad \alpha = \frac{UL}{D}.$$

5. A solution slab has an impermeable wall at $x = 0$, and the solute concentration is kept at c_1 for $x = L$. Initially the concentration has the uniform value c_0. Using the Fourier series

$$\frac{4}{\pi} \sum_{k=0}^{\infty} (-1)^k (2k+1)^{-1} \cos \left(\frac{(2k+1)\pi x}{2L} \right) = 1, \quad 0 \le x < L,$$

show that

$$\frac{c(0,t) - c_1}{c_0 - c_1} = \frac{4}{\pi} \sum_{k=0}^{\infty} \frac{(-1)^k}{2k+1} \exp \left\{ -(2k+1)^2 \pi^2 Dt/4L^2 \right\}.$$

(Hint: use separation of variables on $c(x,t) - c_1$.)

6. Consider a solvent of half-infinite extent in the region $x > 0$. Assume that, in one dimension, a unit mass of solute is placed at the position x_0 at $t = 0$. Find the concentration of solute $c(x,t)$ in the two following cases:

(i) There is an impermeable wall at $x = 0$. Give the value of c at the wall for time t.

(ii) The solute molecules adhere to the wall on contact (i.e., $c(0,t) = 0$). Give the value of flux of solute at $x = 0$, for time t.

7. A spherical cell of radius a is taking in a nutrient from its surroundings and metabolizing it. We assume that this happens instantaneously as each molecule reaches the cell membrane, so that the concentration of nutrient

in the solution at $r = a$ is zero, for $t > 0$. Initially the concentration is c_0 for $r > a$. Solve the differential equation in spherical coordinates, for $c(r, t)$:

$$\frac{\partial c}{\partial t} = D \frac{1}{r^2} \frac{\partial}{\partial r} \left(r^2 \frac{\partial c}{\partial r} \right),$$

and hence find the total flux of nutrient into the cell (area of surface × flux per unit area), at time t. What is the steady-state value of this flux?
(Hint: put $c(r, t) = c_0 + r^{-1} u(r, t)$, substitute into the diffusion equation, and show that u satisfies

$$\frac{\partial u}{\partial t} = D \frac{\partial^2 u}{\partial r^2}$$

with

$$u(r, 0) = 0 \qquad \text{for } r > a,$$
$$u(a, t) = -ac_0 \qquad \text{for } t > 0.$$

Find a combination of the error function and a constant term to satisfy these requirements. To find the flux you will need to use

$$\frac{d}{dz} (\text{erf}(z)) = \frac{2}{\sqrt{\pi}} \exp(-z^2).$$

8. In the nitrous oxide method for determining cerebral blood flow, nitrous oxide is inhaled, and the concentrations of nitrous oxide in the blood entering the brain, $c_a(t)$, and leaving the brain, $c_v(t)$, are measured continuously so that their integrals from 0 to t can be calculated. Use the equation of conservation for the amount $m(t)$ of nitrous oxide in the brain to obtain the result

$$m(t) - m(0) = Q \int_0^t \{c_a(\tau) - c_v(\tau)\} \, d\tau,$$

where Q is the volume rate of blood flow through the brain.

As t tends to infinity, $c_a(t)$ and $c_v(t)$ both tend to c_0, which is also the final concentration of nitrous oxide in the brain tissue. Also the integral of $c_a(\tau) - c_v(\tau)$ from 0 to ∞ has a finite value.

Writing V_B for volume of brain, M_B for mass of brain, $m_0 = c_0 V_B / M_B$, (the mass of nitrous oxide per unit mass of brain at equilibrium), and using $m(0) = 0$, $m(\infty) = V_B c$, show that

$$\frac{Q}{M_B} = m_0 \Big/ \int_0^\infty \{c_a(\tau) - c_v(\tau)\} \, d\tau.$$

9. Using the method of separation of variables, show that the solution of the two-dimensional diffusion equation

$$\frac{\partial c}{\partial t} = D\left(\frac{\partial^2 c}{\partial x^2} + \frac{\partial^2 c}{\partial y^2}\right)$$

inside the rectangle $x = 0$, $x = a$, $y = 0$, $y = b$ can be expressed in the following forms:

(i)

$$c(x,y,t) = \sum_{m=1}^{\infty}\sum_{n=0}^{\infty} A_{mn} \cos\left(\frac{m\pi x}{a}\right) \cos\left(\frac{n\pi y}{b}\right) e^{-\left(\frac{m^2}{a^2}+\frac{n^2}{b^2}\right)\pi^2 Dt},$$

if the boundary conditions are

$$\frac{\partial c}{\partial x} = 0 \quad \text{when } x = 0, a,$$

$$\frac{\partial c}{\partial y} = 0 \quad \text{when } y = 0, b, \quad \text{for all } t.$$

(ii)

$$c(x,y,t) = \sum_{m=1}^{\infty}\sum_{n=0}^{\infty} A_{mn} \sin\left(\frac{m\pi x}{a}\right) \sin\left(\frac{n\pi y}{b}\right) e^{-\left(\frac{m^2}{a^2}+\frac{n^2}{b^2}\right)\pi^2 Dt},$$

if the boundary conditions are

$$c(x,y,t) = 0$$

on the boundary of the rectangle for all t.

Determine the values of A_{mn} in both cases if the initial concentration at all points inside the rectangle is a constant.

10. Show that the one-dimensional convection–diffusion equation

$$\frac{\partial c}{\partial t} + u\frac{\partial c}{\partial x} = D\frac{\partial^2 c}{\partial x^2}$$

can be transformed into

$$\frac{\partial c}{\partial \eta} = D\frac{\partial^2 c}{\partial \xi^2}$$

by the substitutions $\xi = x - ut$, $\eta = t$. Hence, obtain a solution by the method of separation of variables.

3

Population Biology

3.1. Simple Interacting Populations

Population biology or mathematical ecology deals with the increase and fluctuations of populations (for example, plant population, animal population, or other organic population). The mathematical study of the problems in ecology is not of recent origin. In fact, Lotka (1924) and Volterra (1926) were early pioneers developing foundation work in this field. The Lotka–Volterra model of an ecosystem is now a classical model in mathematical ecology.

In this chapter we begin with the problem of two competing species in the same habitat. For example, there is competition between goats and sheep for grass as their food.

If $x = x(t)$ represents the number of first species present at time t and $y = y(t)$ represents the number of second species present at time t, the differential equations describing the competition between the two species are

$$\frac{dx}{dt} = x\left[\alpha - (\gamma_{11}x + \gamma_{12}y)\right],$$
$$\frac{dy}{dt} = y\left[\beta - (\gamma_{21}x + \gamma_{22}y)\right] \tag{3.1}$$

where the coefficients α, β and γ_{ij} are all positive constants. Such equations are commonly referred to as the Lotka–Volterra equations.

An explanation and justification of the Lotka–Volterra competition is given in the following lines.

Here, γ_{12} and γ_{21}, represent the effects of competition between the two species. Consider first the case $\gamma_{12} = \gamma_{21} = 0$. Then the system (3.1) becomes

$$\frac{dx}{dt} = \alpha x - \gamma_{11}x^2,$$
$$\frac{dy}{dt} = \beta y - \gamma_{22}y^2. \tag{3.2}$$

Hence, in the absence of competition, the differential equations are uncoupled so that each species develops independently of the other (which is exactly as expected). Consider now just the x-species. The constant α is known as the natural growth rate of the species. Thus, when the number of species x present is sufficiently small, they multiply and their numbers increase as if

$$\frac{dx}{dt} = \alpha x \tag{3.3}$$

so that

$$x(t) = Ke^{\alpha t}, \tag{3.4}$$

which implies an exponential growth of population. Equation (3.4) has also been called the law of Malthus. Obviously, this result cannot be true for all time t, even in the absence of competition. Eventually there will be an over-population of the species x and there will be insufficient food to provide for the entire population, resulting in death through starvation. Hence the term $-\gamma_{11}x^2$ is included to represent a self limiting growth of the speices. Therefore, the equation governing a single species habitat is

$$\frac{dx}{dt} = \alpha x - \gamma_{11}x^2, \tag{3.5}$$

which is known as the logistic equation. The solution to the logistic equation is easily found to be

$$\frac{x}{1 - (\gamma_{11}/\alpha)x} = Le^{\alpha t}, \tag{3.6}$$

where L is a constant. Clearly, if x is small, the above solution reduces to the previous solution (3.4). Obtaining x explicitly as a function of t from equation (3.6) gives the relationship

$$x = \frac{Le^{\alpha t}}{1 + (\gamma_{11}/\alpha)Le^{\alpha t}} \ \rightarrow \ \frac{\alpha}{\gamma_{11}} \ \text{as } t \rightarrow \infty. \qquad (3.7)$$

That is to say that the population is limited to the value α/γ_{11}, as shown in Figure 3.1.

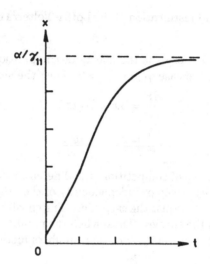

FIGURE 3.1. The solution to the logistic equation approaches asymptotically to α/γ_{11} as $t \rightarrow \infty$

The effect of competition between species x and y is modelled by the addition of the term $-\gamma_{12}xy$ which represents the reduction due to x, y competition. Hence the final equation is

$$\frac{dx}{dt} = \alpha x - \gamma_{11}x^2 - \gamma_{12}xy. \qquad (3.8)$$

The term $-\gamma_{12}xy$ is an empirical correction similar to the law of mass action, in which the competition reduces the growth of each species, at a rate proportional to the product of their amounts.

Another way to view the coupled differential equations (3.1) is to say that the factors $[\alpha - (\gamma_{11}x + \gamma_{12}y)]$ and $[\beta - (\gamma_{21}x + \gamma_{22}y)]$ represent the amount of food available to each species. The rate of growth will then be proportional to the number present capable of reproduction times the food supply. For example, the factor $[\alpha - (\gamma_{11}x + \gamma_{12}y)]$ represents the food available to x: an amount α will be present when there are no eaters at all and this amount is reduced by terms $\gamma_{11}x$ and $\gamma_{12}y$ due to the eating of food by both competing species. In general the two factors are not identical because the two species will eat at different rates. If the factors are identical, we can easily solve the system in (3.1) by division and deduce that y is proportional to x. In this case the two species will be indistinguishable and we can treat the habitat as a single-species one!

The system (3.1) is a special case of systems of the more general form

$$\frac{dx}{dt} = f(x,y),$$
$$\frac{dy}{dt} = g(x,y). \tag{3.9}$$

In system (3.1), f and g are simple quadratic functions of x and y. A more general system than that of (3.9) would be if we consider f and g as functions of t also, i.e.,

$$f = f(x,y,t) \quad \text{and} \quad g = (x,y,t) \tag{3.10}$$

which arises when we are to study systems whose properties change with time. For example, there will be variations in food supply and birth rates due to seasonal climatic changes. However, in this chapter, we neglect such variations by assuming that the time span we are dealing with is sufficiently small to ensure the parameters are constant. Such a system is referred to as an *autonomous* one. Some theorems from the theory of differential equations imply that system (3.10) has a unique solution with initial conditions $(x,y) = (x_0,y_0)$ at time $t = 0$ for sufficiently smooth functions $f(x,y)$ and $g(x,y)$. For example, a sufficiently smooth function in this case is one possessing a Taylor series expansion about (x_0,y_0), that is,

$$f(x,y) = f(x_0,y_0) + (x - x_0)\frac{\partial f}{\partial x}\Big|_{x_0,y_0} + (y - y_0)\frac{\partial f}{\partial y}\Big|_{x_0,y_0} + \cdots. \tag{3.11}$$

Certain quadratic functions like those employed in system (3.1) are entirely acceptable. Note that system (3.9) can be reduced to a single first-order ordinary differential equation

$$\frac{dy}{dx} = \frac{g(x,y)}{f(x,y)} \tag{3.12}$$

by division. In certain cases, the above differential equation (3.12) can be solved, but the system (3.1) is not analytically solvable.

3.2. Equilibrium Analysis

The ecologist is often interested in possible uniform states where the two species coexist. That is to say, where the populations are in equilibrium, with $dx/dt = dy/dt = 0$. Hence, from system (3.9) we set

$$f(x,y) = 0 \quad \text{and} \quad g(x,y) = 0. \tag{3.13}$$

This is a set of equations which may have solutions

$$(x,y) = (x_0, y_0). \tag{3.14}$$

Such points are known as *critical points* or *equilibrium points* of the system. In general, there may be no, one, two or many such points. For example, setting the right of system (3.1) to zero we obtain

$$\text{either } x = 0 \text{ or } \gamma_{11}x + \gamma_{12}y = \alpha, \tag{3.15a}$$

and

$$\text{either } y = 0 \text{ or } \gamma_{21}x + \gamma_{22}y = \beta. \tag{3.15b}$$

An obvious critical point is the origin $(x,y) = (0,0)$. This is the trivial case in which both species are extinct and there is no possibility of growth. Ecologically this is not a very interesting case! We now look for other possibilities.

Suppose $x = 0$, then either $y = 0$ (considered above) or $\gamma_{22}y = \beta$. Hence, for $\gamma_{22} \neq 0$, $y = \beta/\gamma_{22}$. So we find that $(0, \beta/\gamma_{22})$ is an equilibrium point. This means that species x is extinct and y has reached its equilibrium population β/γ_{22}.

In the same manner, if $y = 0$, then either $x = 0$, as discussed before, or $\gamma_{11}x = \alpha$. So another equilibrium point is $(\alpha/\gamma_{11}, 0)$. Here, species y is extinct and x has thrived and reached its equilibrium population. Obviously, the ecologist is interested in deciding which case is likely to eventuate.

What of the case of the two species coexisting? That is, neither x nor y being extinct. We obtain

$$\begin{aligned} \gamma_{11}x + \gamma_{12}y &= \alpha, \\ \gamma_{21}x + \gamma_{22}y &= \beta \end{aligned} \tag{3.16}$$

or $\Gamma \underset{\sim}{x} = \underset{\sim}{\alpha}$ where

$$\Gamma = \begin{pmatrix} \gamma_{11} & \gamma_{12} \\ \gamma_{21} & \gamma_{22} \end{pmatrix}, \quad \underset{\sim}{\alpha} = \begin{pmatrix} \alpha \\ \beta \end{pmatrix}. \tag{3.17}$$

This has a nonzero unique solution $(x,y) = (x_c, y_c)$ if and only if $\det(\Gamma) \neq 0$. If $\det(\Gamma) = 0$, then there is no critical point (except for special α and β in which case we may obtain a line of critical points, a case which is excluded here).

So for $\det(\Gamma) = 0$, the only equilibrium points are those with at least one species dead. If $\det(\Gamma) \neq 0$ and $x_c, y_c > 0$, we have a possible equilibrium with coexistence of both species. Such a situation is of importance to ecologists.

3.3. Behaviour of Trajectories in the Neighbourhood of Critical Points

Assume that $f(x,y)$ and $g(x,y)$ are both analytic functions near the isolated critical point (x_c, y_c) so that they possess Taylor series expansions about that point as

$$f(x,y) = f(x_c,y_c) + (x - x_c)f_x(x_c,y_c) + (y - y_c)f_y(x_c,y_c)$$
$$+ \frac{1}{2}(x - x_c)^2 f_{xx}(x_c,y_c) + \frac{1}{2}(y - y_c)^2 f_{yy}(x_c,y_c)$$
$$+ (x - x_c)(y - y_c)f_{xy}(x_c,y_c) + \cdots,$$

$$g(x,y) = g(x_c,y_c) + (x - x_c)g_x(x_c,y_c) + (y - y_c)g_y(x_c,y_c)$$
$$+ \frac{1}{2}(x - x_c)^2 g_{xx}(x_c,y_c) + \frac{1}{2}(y - y_c)^2 g_{yy}(x_c,y_c)$$
$$+ (x - x_c)(y - y_c)g_{xy}(x_c,y_c) + \cdots$$

where

$$f_x = \frac{\partial f}{\partial x}, \quad f_y = \frac{\partial f}{\partial y}, \quad f_{xx} = \frac{\partial^2 f}{\partial x^2}, \quad \text{etc.}$$

But from the definition of critical points, $f(x_c,y_c) = g(x_c,y_c) = 0$ so that

$$f(x,y) = (x - x_c)f_x(x_c,y_c) + (y - y_c)f_y(x_c,y_c) + \cdots,$$
$$g(x,y) = (x - x_c)g_x(x_c,y_c) + (y - y_c)g_y(x_c,y_c) + \cdots.$$

Denoting

$$a_{11} = f_x(x_c,y_c), \quad a_{12} = f_y(x_c,y_c),$$
$$a_{21} = g_x(x_c,y_c), \quad a_{22} = g_y(x_c,y_c)$$

and putting $X = x - x_c$, $Y = y - y_c$, so that (X,Y) are coordinates referred to the critical point (x_c, y_c) as the centre, and setting $R = \sqrt{X^2 + Y^2}$, we finally obtain

$$f = a_{11}X + a_{12}Y + O(R^2),$$
$$g = a_{21}X + a_{22}Y + O(R^2) \tag{3.18}$$

where $O(R^2)$ means of the order of R^2. As R tends to zero, that is as $(X,Y) \to 0$, we deduce that $(x,y) \to (x_c,y_c)$. Hence, near the critical point, the general system of differential equations (3.9) reduces to the pair of first-order differential equations as

$$\frac{dX}{dt} = a_{11}X + a_{12}Y,$$

$$\frac{dY}{dt} = a_{21}X + a_{22}Y \tag{3.19}$$

with error $O(R^2)$. In an equivalent matrix form we have

$$\frac{d\underset{\sim}{x}}{dt} = A\underset{\sim}{x},$$

where

$$\underset{\sim}{x} = \begin{pmatrix} X \\ Y \end{pmatrix}, \qquad A = \begin{pmatrix} a_{11} & a_{12} \\ a_{21} & a_{22} \end{pmatrix}.$$

Thus, for every nonlinear system (3.9), we may locally linearize the system (3.19), where the coefficients a_{ij} are just the values of the particular derivatives at the critical point (x_c, y_c).

Some examples of local linearization are as follows:

(i) Consider the case near the origin $(0,0)$. We have $X = x$ and $Y = y$. Also $f(x,y) = x[\alpha - (\gamma_{11}x + \gamma_{12}y)]$, $g(x,y) = y[\beta - (\gamma_{21}x + \gamma_{22}y)]$, so $a_{11} = f_x(0,0) = \alpha$, $a_{22} = g_y(0,0) = \beta$, $a_{12} = f_y(0,0) = 0$, $a_{21} = g_x(0,0) = 0$ and

$$\frac{dX}{dt} = \alpha X + O(R^2),$$

$$\frac{dY}{dt} = \beta Y + O(R^2). \tag{3.20}$$

In this example,

$$A = \begin{pmatrix} \alpha & 0 \\ 0 & \beta \end{pmatrix}.$$

The solution of the above linear approximation is simply

$$x = X = c_1 e^{\alpha t},$$

$$y = Y = c_2 e^{\beta t}, \tag{3.21}$$

where c_1 and c_2 are constants. Thus, both species grow at their free rates, unhindered by either competition or their own numbers. This is necessarily the way small populations behave in an abundant food supply.

(ii) Next consider near the critical point $(x_c, y_c) = (0, \beta/\gamma_{22})$. By putting $x_c = 0$, $y_c = \beta/\gamma_{22}$, we have $X = x$, $Y = y - \beta/\gamma_{22}$ and hence

$$\frac{dX}{dt} = X[\alpha - (\gamma_{11}X + \gamma_{12}(Y + \beta/\gamma_{22}))],$$

$$\frac{dY}{dt} = (Y + \beta/\gamma_{22})[\beta - (\gamma_{21}X + \gamma_{22}(Y + \beta/\gamma_{22}))].$$

(3.22)

By neglecting all quadratic terms (because X and Y are small) the above differential equations reduce to

$$\frac{dX}{dt} = \left(\alpha - \frac{\gamma_{12}}{\gamma_{22}}\beta\right)X + O(R^2),$$

$$\frac{dY}{dt} = -\frac{\gamma_{21}}{\gamma_{22}}\beta X - \beta Y + O(R^2).$$

(3.23)

Here we have

$$A = \begin{pmatrix} \lambda & 0 \\ -\frac{\gamma_{21}}{\gamma_{22}}\beta & -\beta \end{pmatrix}, \qquad \lambda = \alpha - \frac{\gamma_{12}}{\gamma_{22}}\beta.$$

The elements of the matrix A may also be calculated by partial differentiation and provide us with a useful check:

$$a_{11} = f_x(x_c, y_c) = \alpha - \frac{\gamma_{12}}{\gamma_{22}}\beta,$$

$$a_{12} = f_y(x_c, y_c) = 0,$$

$$a_{21} = g_x(x_c, y_c) = -\frac{\gamma_{21}}{\gamma_{22}}\beta,$$

$$a_{22} = g_y(x_c, y_c) = -\beta.$$

(3.24)

The solution to the reduced differential equation

$$\frac{dX}{dt} = \lambda X$$

is simply

$$X(t) = C_1 e^{\lambda t}.$$

(3.25)

Hence by substitution of this, we obtain

$$\frac{dY}{dt} + \beta Y = -\frac{\gamma_{21}}{\gamma_{22}}\beta C_1 e^{\lambda t},$$

the solution of which is given by

$$Y(t) = C_2 e^{-\beta t} - \frac{\beta\gamma_{21}C_1}{(\lambda + \beta)\gamma_{22}}e^{\lambda t},$$

(3.26)

where C_1 and C_2 are arbitrary constants.

If $\lambda < 0$, then both $X \to 0$ and $Y \to 0$ as $t \to \infty$. Hence, $(x, y) \to (x_c, y_c) = (0, \beta/\gamma_{22})$. That is to say, the critical point $(0, \beta/\gamma_{22})$ is locally stable. In this case, species x becomes extinct and y reaches equilibrium. If $\lambda > 0$, then both $X \to \infty$ and $Y \to \infty$ as $t \to \infty$. Hence, this critical point is never approached as $t \to \infty$. In this case, species x does not become extinct.

Note that the condition $\lambda < 0$ for extinction of x implies that $\alpha < \beta\gamma_{12}/\gamma_{22}$. That is, if there is an insufficient initial growth rate α of x compared to the initial growth rate β of y, then species x dies out. Thus, if $\lambda > 0$ then with initial values (x_0, y_0) near the critical point $(0, \beta/\gamma_{22})$, the values of (x, y) move away from $(0, \beta/\gamma_{22})$.

(iii) For system (3.1) near the critical point $(\alpha/\gamma_{11}, 0)$, we may similarly deduce that

$$A = \begin{pmatrix} -\alpha & -\alpha\gamma_{12}/\gamma_{11} \\ 0 & \mu \end{pmatrix}, \qquad \mu = \beta - \frac{\gamma_{21}}{\gamma_{11}}\alpha.$$

Hence, we may derive the same type of conclusion as in (ii) above, namely that if $\mu < 0$, species y will become extinct. The condition $\beta < \alpha\gamma_{21}/\gamma_{11}$ ensures that the growth rate of y compared to x is too low and y dies out.

(iv) Now suppose for system (3.1) we are near the positive critical point (x_c, y_c). Then, because $\gamma_{11}x_c + \gamma_{12}y_c = \alpha$ and $\gamma_{21}x_c + \gamma_{22}y_c = \beta$, it is easily shown that $a_{11} = -x_c\gamma_{11}$, $a_{12} = -x_c\gamma_{12}$, $a_{21} = -y_c\gamma_{21}$, and $a_{22} = -y_c\gamma_{22}$, so that

$$A = \begin{pmatrix} -x_c\gamma_{11} & -x_c\gamma_{12} \\ -y_c\gamma_{21} & -y_c\gamma_{22} \end{pmatrix}. \tag{3.27}$$

In this case, the general system

$$\frac{d\underset{\sim}{x}}{dt} = \underset{\sim}{\dot{x}} = A\underset{\sim}{x} \tag{3.28}$$

cannot be easily solved.

However, the general solution of the linear system $\underset{\sim}{\dot{x}} = A\underset{\sim}{x}$ can be found using the matrix method, by guessing a solution of the form

$$\underset{\sim}{x} = \underset{\sim}{c}e^{\lambda t} \tag{3.29}$$

where $\underset{\sim}{c}$ is a constant vector and λ is a scalar. Substituting in (3.28) and cancelling $e^{\lambda t}$, we find that

$$A\underset{\sim}{c} = \lambda\underset{\sim}{c}. \tag{3.30}$$

Hence, λ is an eigenvalue of the matrix A and c is the corresponding eigenvector. In general, unless A is symmetric, even though the elements of A are real, λ will be a complex number.

Since A is a 2×2 matrix there will be two eigenvalues λ_1 and λ_2 and two corresponding eigenvectors c_1 and c_2, which are linearly independent if $\lambda_1 \neq \lambda_2$. Therefore, the general solution to the linear system $\dot{\underset{\sim}{x}} = A\underset{\sim}{x}$ is

$$\underset{\sim}{x} = \underset{\sim}{c_1}e^{\lambda_1 t} + \underset{\sim}{c_2}e^{\lambda_2 t}. \tag{3.31}$$

3.4. Stability and Classification of Equilibrium Points

It is convenient to plot $(x(t), y(t))$ in the (x, y) plane, with t as a parameter. Then every solution of the differential equation system (3.9) describes a curve in the (x, y) plane as t varies, and is known as a *trajectory*. If the trajectory through (x_0, y_0) consists entirely of the point (x_0, y_0), then this point is an equilibrium point (i.e., a critical point).

An equilibrium point (x_c, y_c) is said to be stable if both the exponentials in equation (3.31) are decreasing. Equivalently, we require that $\Re(\lambda_1) < 0$ and $\Re(\lambda_2) < 0$. This means that both the exponentials in (3.31) tend to zero as t tends to infinity. The critical point is unstable if either (or both) exponential increases so that we require $\Re(\lambda_1) > 0$ and/or $\Re(\lambda_2) > 0$. This means that in (3.29) x tends to infinity as t tends to infinity. The borderline case is said to be marginally stable if it is not unstable but one of the λ's has a zero real part.

An equilibrium point surrounded by a family of closed trajectories, but not approached by any trajectory, is called a *centre*. If all the trajectories approach an equilibrium point, the point is called a *stable node* but, if all the trajectories diverge from an equilibrium point, then the point is called an *unstable node*. If the trajectories have two asymptotes passing through the equilibrium point, then the point is called an unstable *saddle point*. If the trajectories are spirals converging to the equilibrium, then the point is called a *stable focus*. If these spirals diverge from the equilibrium point, then the point is called an *unstable focus*. These cases are all shown in Figure 3.2.

If however the trajectory through the equilibrium point returns to the equilibrium, then the resulting simple closed curve is called a *limit cycle* and is the phase-plane path of a periodic solution.

Consider the differential equation system (3.9) given by

$$\frac{dx}{dt} = x + y - x(x^2 + y^2),$$
$$\frac{dy}{dt} = -x + y - y(x^2 + y^2).$$

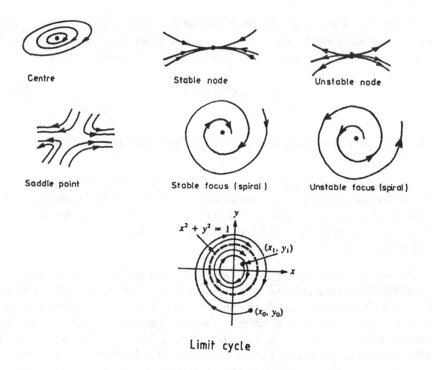

FIGURE 3.2. Classification of equilibrium points

It can be seen that (0,0) is the only equilibrium point of the system. Yet any trajectory starting on the unit circle $x^2 + y^2 = 1$ will traverse the unit circle in a periodic manner because in that case $dy/dx = -x/y$. Moreover, if a trajectory starts inside the circle (provided it does not start at the origin) it will spiral outward asymptotically, getting closer and closer to the circular path as t tends to infinity. Similarly, if the trajectory starts outside the circular region, it will spiral inward and finally approach the circular path asymptotically. The solution $x^2 + y^2 = 1$ is called a limit cycle. The trajectory is shown in Figure 3.2. Thus, if a population behaviour for two competing species is modelled by the above system, it can be concluded that the population levels will eventually be periodic.

3.5. Relationship Between Eigenvalues and Critical Points

Note that critical points are classified by the behaviour of the eigenvalues λ_1 and λ_2 of the matrix A. Suppose the matrix A is real. The specific eigenvalue equation is $Ac = \lambda c$ so that we have the determinant

$$\begin{vmatrix} a_{11} - \lambda & a_{12} \\ a_{21} & a_{22} - \lambda \end{vmatrix} = 0. \tag{3.32}$$

Hence,

$$\lambda^2 - (a_{11} + a_{22})\lambda + (a_{11}a_{22} - a_{21}a_{12}) = 0. \tag{3.33}$$

Now, for the roots λ_1 and λ_2 to the above equation to be real, we require

$$(a_{11} + a_{22})^2 \geq 4a_{11}a_{22} - 4a_{21}a_{12}$$

or

$$(a_{11} - a_{22})^2 \geq -4a_{21}a_{12}. \tag{3.34}$$

This condition for real λ's is always true if the off-diagonal elements are of the same sign.

Once we have established the nature of the roots (i.e., real or complex) we can demonstrate the classification of critical points. Let us consider the following cases:

(a) λ_1, λ_2 are real and negative

Suppose $\lambda_2 < \lambda_1 < 0$ (a similar argument holds if $\lambda_2 = \lambda_1 < 0$). The general solution is

$$x = c_1 e^{\lambda_1 t} + c_2 e^{\lambda_2 t}, \tag{3.35}$$

where c_1 and c_2 are linearly independent eigenvectors of A corresponding to eigenvalues λ_1, λ_2. Clearly as $t \to \infty$, $x \to 0$. But we know that for large t, $e^{\lambda_2 t} \ll e^{\lambda_1 t}$ and hence

$$x \approx c_1 e^{\lambda_1 t}. \tag{3.36}$$

Let p_j, q_j be the components of eigenvector c_j, then

$$c_1 = \begin{pmatrix} p_1 \\ q_1 \end{pmatrix}.$$

Therefore

$$x \to p_1 e^{\lambda_1 t},$$
$$y \to q_1 e^{\lambda_1 t},$$

so

$$\frac{y}{x} \to \frac{q_1}{p_1}, \quad \text{a constant, as } t \to \infty,$$

which is the same for all trajectories very near this equilibrium point. That is, as $t \to \infty$, all trajectories approach the critical point in question along the unique direction with gradient q_1/p_1. In order to sketch the trajectory, we first calculate the eigenvalues λ_1 and λ_2. The eigenvector c_1 corresponding to the minimum λ is then evaluated and from its components $\underset{\sim}{c}$ we obtain the slope of the limiting line. Clearly in this case we get a *stable node* (see Figure 3.3 for trajectories).

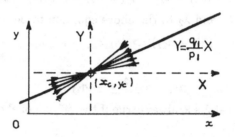

FIGURE 3.3. Stable node trajectories

(b) λ_1, λ_2 are real and positive

Suppose $\lambda_2 > \lambda_1 > 0$ (a similar argument holds if $\lambda_2 = \lambda_1 > 0$). This case is similar to (a) above with t (time) taking the opposite sign. That is, we now consider the limit as $t \to -\infty$, which means that we are considering where trajectories came from. Here, as $t \to -\infty$, $\underset{\sim}{x} \to 0$, but since $\lambda_2 > \lambda_1 > 0$ then $e^{\lambda_2 t} \to 0$ faster than $e^{\lambda_1 t} \to 0$ as $t \to -\infty$. Therefore $\underset{\sim}{x} \approx \underset{\sim}{c_1} e^{\lambda_1 t}$ as $t \to -\infty$ or

$$
\begin{aligned}
x &\to p_1 e^{\lambda_1 t}, \\
y &\to q_1 e^{\lambda_1 t}.
\end{aligned}
\tag{3.37}
$$

So the trajectories all originate from the equilibrium point along the unique straight line of slope

$$
\frac{y}{x} = \frac{q_1}{p_1}.
$$

This is an *unstable node* and the trajectories are shown in Figure 3.4.

For example consider the system in the neighbourhood of the equilibrium point $(0,0)$. Then substitution of $x = X$, $y = Y$ and linearizing gives

$$
\begin{aligned}
\frac{dX}{dt} &= \alpha X, \\
\frac{dY}{dt} &= \beta Y,
\end{aligned}
\tag{3.38}
$$

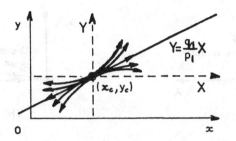

FIGURE 3.4. Unstable node trajectories

so

$$A = \begin{pmatrix} \alpha & 0 \\ 0 & \beta \end{pmatrix}.$$

Clearly, the eigenvalues of A are $\lambda = \alpha, \beta$ which are both real and positive, so $(0,0)$ is an *unstable node*. Suppose $\alpha > \beta > 0$, then taking $\lambda_1 = \beta$, $\lambda_2 = \alpha$, we obtain the eigenvector

$$\underset{\sim}{c_1} = \begin{pmatrix} p_1 \\ q_1 \end{pmatrix}$$

by solving $A\underset{\sim}{c_1} = \lambda_1 \underset{\sim}{c_1}$ for $\underset{\sim}{c_1}$, or

$$\begin{pmatrix} \alpha & 0 \\ 0 & \beta \end{pmatrix} \begin{pmatrix} p_1 \\ q_1 \end{pmatrix} = \beta \begin{pmatrix} p_1 \\ q_1 \end{pmatrix}, \tag{3.39}$$

which leads to

$$\alpha p_1 = \beta p_1 \;\to\; p_1 = 0 \text{ as } \alpha \neq \beta,$$
$$\beta q_1 = \beta q_1 \;\to\; q_1, \text{ which is arbitrary.}$$

So a possible $\underset{\sim}{c_1}$ is

$$\underset{\sim}{c_1} = \begin{pmatrix} 0 \\ 1 \end{pmatrix}.$$

The axis of the node is therefore $X/Y = 0/1$, i.e., the Y-axis (see Figure 3.5a).

Alternatively, if $\beta > \alpha > 0$, then $\lambda_1 = \alpha$, $\lambda_2 = \beta$, and

$$\underset{\sim}{c_1} = \begin{pmatrix} 1 \\ 0 \end{pmatrix},$$

so that the axis of the node is $Y/X = 0/1$, i.e., the X-axis (see Figure 3.5b).

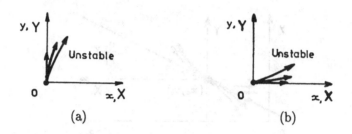

FIGURE 3.5. Unstable node trajectories

(c) λ_1, λ_2 are real and of opposite sign

Suppose $\lambda_1 > 0$, $\lambda_2 < 0$ with corresponding eigenvectors

$$\underset{\sim}{c_1} = \begin{pmatrix} p_1 \\ q_1 \end{pmatrix}, \qquad \underset{\sim}{c_2} = \begin{pmatrix} p_2 \\ q_2 \end{pmatrix}. \tag{3.40}$$

Then

$$\underset{\sim}{X} = \underset{\sim}{c_1} e^{\lambda_1 t} + \underset{\sim}{c_2} e^{\lambda_2 t}. \tag{3.41}$$

As $t \to \infty$,

$$\underset{\sim}{X} \approx \underset{\sim}{c_1} e^{\lambda_1 t}, \qquad \frac{Y}{X} \approx \frac{q_1}{p_1}, \tag{3.42}$$

i.e., the trajectories asymptote along an axis of slope $Y/X = q_1/p_1$. As $t \to -\infty$

$$\underset{\sim}{X} \approx \underset{\sim}{c_2} e^{\lambda_2 t}, \qquad \frac{Y}{X} \approx \frac{q_2}{p_2}, \tag{3.43}$$

i.e., the trajectories asymptote along an axis of slope $Y/X = q_2/p_2$. This is the case of an *unstable saddle point*.

For example consider, after centring the system on (x_c, y_c), that $dX/dt = X$, and $dY/dt = -Y$, so that

$$A = \begin{pmatrix} 1 & 0 \\ 0 & -1 \end{pmatrix}, \quad \lambda_1 = 1, \ \lambda_2 = -1, \ \underset{\sim}{c_1} = \begin{pmatrix} 1 \\ 0 \end{pmatrix}, \ \underset{\sim}{c_2} = \begin{pmatrix} 0 \\ 1 \end{pmatrix},$$

$$\frac{q_1}{p_1} = 0, \quad \frac{q_2}{p_2} = \infty.$$

In this case,

$$X = e^t \to 0 \text{ as } t \to -\infty$$

and

$$Y = e^{-t} \to 0 \text{ as } t \to +\infty$$

These cases are shown in Figure 3.6.

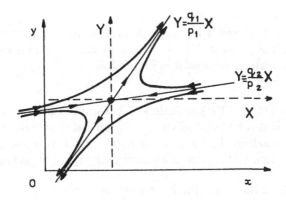

FIGURE 3.6. Saddle point trajectories

(d) λ_1, λ_2 are complex conjugates

We first write λ_1, $\lambda_2 = \eta \pm i\sigma$ and proceed as follows:

(i) $\Re(\lambda) = \eta > 0$

The general solution (3.31) now takes the form

$$X = c_1 e^{(\eta+i\sigma)t} + c_2 e^{(\eta-i\sigma)t}$$

$$\text{or} \quad X = d_1 e^{\eta t} \cos(\sigma t) + d_2 e^{\eta t} \sin(\sigma t), \tag{3.44}$$

where $d_1 = c_1 + c_2$, and $d_2 = i(c_1 - c_2)$. The above set of equations can also be written as

$$X(t) = k_1 e^{\eta t} \cos(\sigma t - \epsilon_1),$$

$$Y(t) = k_2 e^{\eta t} \cos(\sigma t - \epsilon_2), \tag{3.45}$$

where k_1 and k_2 are positive constants. Since $\eta > 0$, then $X(t)$ and $Y(t)$ oscillate with increasing amplitude as t increases, with $X(t) \to 0$, $Y(t) \to 0$ as $t \to -\infty$. Therefore trajectories move away from the equilibrium point in a spiral-like manner, so this is an *unstable spiral point* and is illustrated in Figure 3.7.

Note that the winding direction of the spiral is easily found by examining the original differential equation. For example, if x is very close to zero, and $y > 0$ and the resulting value for dx/dt is greater than zero in this region, then the spiral must be an anti-clockwise one.

(ii) $\Re(\lambda) = \eta < 0$

In this case $X(t)$ and $Y(t)$ oscillate with decreasing amplitude as t increases, with $X(t) \to 0$, $Y(t) \to 0$ as $t \to \infty$. Therefore we have *stable spiral points*, which are shown in Figure 3.8.

(iii) $\Re(\lambda) = \eta = 0$

In this case $X(t)$ and $Y(t)$ oscillate with constant amplitudes as t increases, so $X(t)$ oscillates between $\pm k_1$ and $Y(t)$ oscillates between $\pm k_2$ in closed curves surrounding the equilibrium point, which is a *marginally stable centre* (see Figure 3.9). Note the winding direction can be found as above.

An example of this is the predator–prey equations

$$\frac{dx}{dt} = x(\alpha - \gamma_1 y),$$
$$\frac{dy}{dt} = -y(\beta - \gamma_2 x), \tag{3.46}$$

which is a certain form of the Lotka–Volterra equations. Here, species x is the prey and y is the predator. The critical point of interest is $(x_c, y_c) = (\beta/\gamma_2, \alpha/\gamma_1)$ obtained by setting $\dot{x} = \dot{y} = 0$.

As before, we define the new variables X and Y as

$$X = x - \frac{\beta}{\gamma_2}, \qquad Y = y - \frac{\alpha}{\gamma_1}.$$

By substituting into the predator–prey equations and linearizing by neglecting all quadratic terms we obtain

$$\dot{X} = \frac{\beta}{\gamma_2}(-\gamma_1 Y),$$
$$\dot{Y} = -\frac{\alpha}{\gamma_1}(-\gamma_2 X). \tag{3.47}$$

The above linear system can be solved by eliminating Y from the first of the equations:

$$\ddot{X} = -\frac{\gamma_1}{\gamma_2}\beta\dot{Y} = \left(-\frac{\gamma_1}{\gamma_2}\beta\right)\left(\frac{\gamma_2}{\gamma_1}\alpha\right)X = -\alpha\beta X. \tag{3.48}$$

Hence, the simple differential equation we must solve is

$$\ddot{X} + (\alpha\beta)X = 0 \tag{3.49}$$

whose solution is

$$X = A\cos\sqrt{\alpha\beta}t + B\sin\sqrt{\alpha\beta}t. \tag{3.50}$$

By substitution we find that

$$Y = -\frac{\gamma_2}{\gamma_1 \beta}\dot{X} = \frac{\gamma_2}{\gamma_1 \beta}\sqrt{\alpha\beta}\left[A\sin\sqrt{\alpha\beta}t - B\cos\sqrt{\alpha\beta}t\right]. \qquad (3.51)$$

We can easily see that the solutions oscillate about the critical point $(\beta/\gamma_2, \alpha/\gamma_1)$ and the fundamental natural frequency of the system is $\sqrt{\alpha\beta}$.

The solutions for X and Y may not at first glance appear to be ellipses. However, by introducing some phase angle δ such that $\tan\delta = B/A$ we can write

$$X = \sqrt{A^2 + B^2}\cos\left(\sqrt{\alpha\beta}t - \delta\right),$$

$$Y = \sqrt{A^2 + B^2}\frac{\gamma_2}{\gamma_1}\sqrt{\frac{\alpha}{\beta}}\sin\left(\sqrt{\alpha\beta}t - \delta\right), \qquad (3.52)$$

where $\cos\delta = A/\sqrt{A^2 + B^2}$ and $\sin\delta = B/\sqrt{A^2 + B^2}$. Hence we can obtain, by setting $a = \sqrt{A^2 + B^2}$ and $b = \sqrt{A^2 + B^2}(\gamma_2/\gamma_1)\sqrt{\alpha/\beta}$, the familiar equation of the ellipse

$$\frac{X^2}{a^2} + \frac{Y^2}{b^2} = 1. \qquad (3.53)$$

Note that the ratio b/a is the same for all trajectories.

(Anti-clockwise)   (Clockwise)

FIGURE 3.7. Unstable spiral point

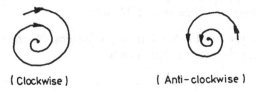

(Clockwise)   (Anti-clockwise)

FIGURE 3.8. Stable spiral point

Whenever we have centre type trajectories, the number of species $x(t)$ and $y(t)$ will always oscillate with time. This has an interesting application to ecological systems. If there are large numbers of prey (species x) present then the predators will feed well and for a period will prosper in numbers.

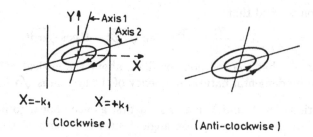

(Clockwise)   (Anti-clockwise)

FIGURE 3.9. Marginally stable point

However, the amount of prey will eventually become too small to support all the predators so the latter will experience a drop in population. When this population becomes small enough, the prey can enjoy a period of relative safety and expand in numbers and so the cycle repeats itself continually.

From the solution of the system, we see that species x and y are always $90°$ out of phase. This represents one quarter of a period or $\pi/2\sqrt{\alpha\beta}$ units of time.

Now that we have discussed the classification of the trajectories and their critical points we may discuss the competing species problem.

We have already seen that the critical point $(0,0)$ is an unstable node. Consider then the critical point $(0, \beta/\gamma_{22})$. The matrix A under consideration is easily seen to be

$$A = \begin{pmatrix} \lambda_1 & 0 \\ -\frac{\gamma_{21}}{\gamma_{22}}\beta & -\beta \end{pmatrix}$$

where $\lambda_1 = \alpha - \gamma_{12}\beta/\gamma_{22}$ and $\lambda_2 = -\beta$. Obviously $\lambda_2 < 0$ and for $\lambda_1 < 0$ we require $\alpha < \gamma_{12}\beta/\gamma_{22}$, resulting in a stable node.

Now consider the local behaviour of the system. We have already found in (3.25) and (3.26) the exact solution to be

$$X = C_1 e^{\lambda_1 t},$$

$$Y = C_2 e^{-\beta t} - \frac{\beta\gamma_{21}C_1}{(\lambda_1 + \beta)\gamma_{22}} e^{\lambda_1 t}.$$

There are two cases then. If $\lambda_2 = -\beta < \lambda_1$, then $e^{-\beta t}$ goes to zero faster than $e^{\lambda_1 t}$ and hence Y tends to the term in $e^{\lambda_1 t}$. Therefore

$$\frac{Y}{X} \to -\frac{\beta\gamma_{21}}{(\lambda_1 + \beta)\gamma_{22}} < 0.$$

If $\lambda_2 = -\beta > \lambda_1$, then $e^{-\beta t}$ tends to zero slower than $e^{\lambda_1 t}$ so that

$$Y \to C_2 e^{-\beta t} \gg X$$

or in other words, X tends to zero relative to the value of Y.

We thus have that the trajectories approach the critical point either along the y axis (see Figure 3.10a) in which case $\lambda_1 < -\beta$ and $\alpha < (\gamma_{12}/\gamma_{22} - 1)\beta$, a case when x becomes extinct very quickly, or along sloping lines (see Figure 3.10b)

$$Y = -\frac{\beta\gamma_{21}}{(\lambda_1 + \beta)\gamma_{22}} X$$

if $\lambda_1 > -\beta$, giving

$$\left(\frac{\gamma_{12}}{\gamma_{22}} - 1\right)\beta < \alpha < \frac{\gamma_{12}}{\gamma_{22}}\beta.$$

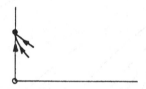

FIGURE 3.10a. Trajectories approaching the critical point along the y-axis

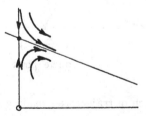

FIGURE 3.10b. Trajectories approaching the critical point along sloping lines

Otherwise if we have the case $\alpha > \gamma_{12}\beta/\gamma_{22}$ so that $\lambda_1 > 0$, an unstable saddle point is obtained because $\lambda_2 = -\beta < 0$. So as time approaches infinity,

$$X \to C_1 e^{\lambda_1 t}, \qquad Y \to -\frac{\beta\gamma_{21}}{(\lambda_1 + \beta)\gamma_{22}} X$$

for $t \to -\infty$, $X \to 0$, $Y \to \infty$.

When discussing the critical point at $(\alpha/\gamma_{11}, 0)$ similar arguments are used to those just discussed.

Suppose, for certain parameter values, we deduce that one of the critical points is stable and the other is unstable. By examining the equations governing the system, suppose also that there is no critical point possible with $x_c > 0$, $y_c > 0$. Then we already have enough information to be able to sketch all the trajectories. For example, consider the case where species x becomes extinct so that $(0, \beta/\gamma_{22})$ is a stable node and $(\alpha/\gamma_{11}, 0)$ is an unstable saddle point (see Figure 3.11).

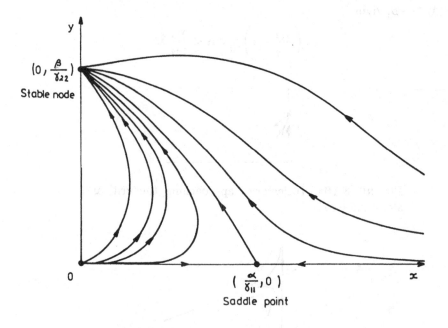

FIGURE 3.11. Sketch of trajectories leading to extinction of x

Clearly, every trajectory leads to the extinction of x, except when there is none of species y to begin with. So if the trajectory begins along the x-axis, then the trajectory will approach the sadddle point.

It is interesting to note that in the case where there is no positive critical point, the question of which species dies out is uniquely determined by the parameters α, β, γ_{ij} of the differential equations and not by the initial conditions (except of course when a species is extinct to begin with).

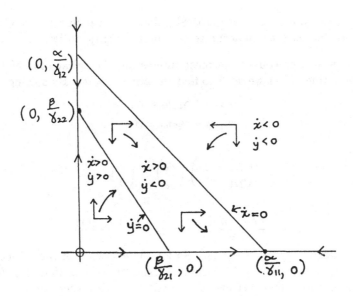

FIGURE 3.12. Division of the first quadrant into 3 distinct regions

The matter of which species survives can be more simply treated by graphical means only. By examining the differential equations

$$\frac{dx}{dt} = x[\alpha - (\gamma_{11}x + \gamma_{12}y)],$$
$$\frac{dy}{dt} = y[\beta - (\gamma_{21}x + \gamma_{22}y)]$$

(3.54)

there should be distinct regions in which dx/dt and dy/dt are greater than and less than zero and distinct lines for the borderline case. By setting dx/dt and dy/dt equal to zero we can draw 2 lines on the (x,y) plane dividing the positive quadrant into 3 distinct regions. We also determine the critical points and for each region we deduce whether dx/dt and dy/dt is positive or negative and draw arrows to indicate whether x and y are increasing or decreasing (see Figure 3.12).

For example, consider the case where

$$\frac{\alpha}{\gamma_{11}} > \frac{\beta}{\gamma_{21}} \quad \text{and} \quad \frac{\alpha}{\gamma_{12}} > \frac{\beta}{\gamma_{22}}$$

where the only two critical points of interest are $(0, \beta/\gamma_{22})$ and $(\alpha/\gamma_{11}, 0)$.

Although the conclusion as to which species survives is the same as deduced by the more rigorous mathematical procedure, the graphical solution tells us nothing about the nature of the critical points (i.e., saddle or node etc.) and does not provide us with an accurate sketch of the trajectories.

There is also of course the most interesting ecological case of the two species coexisting. First we need to find the critical point. By solving

$$\gamma_{11}x_c + \gamma_{12}y_c = \alpha,$$
$$\gamma_{21}x_c + \gamma_{22}y_c = \beta$$

or

$$\begin{pmatrix} \gamma_{11} & \gamma_{12} \\ \gamma_{21} & \gamma_{22} \end{pmatrix} \begin{pmatrix} x_c \\ y_c \end{pmatrix} = \begin{pmatrix} \alpha \\ \beta \end{pmatrix},$$

we obtain

$$x_c = \frac{\alpha\gamma_{22} - \beta\gamma_{12}}{\Delta}, \qquad y_c = \frac{\beta\gamma_{11} - \alpha\gamma_{21}}{\Delta},$$

where $\Delta = \det \gamma = \gamma_{11}\gamma_{22} - \gamma_{12}\gamma_{21}$. So for the existence of the critical point we require $x_c > 0$ and $y_c > 0$. Assume for the time being that $\Delta > 0$. This is a reasonable statement because if we expect stability then the competition terms γ_{12} and γ_{21} would not be large enough to dominate γ_{11} and γ_{22}.

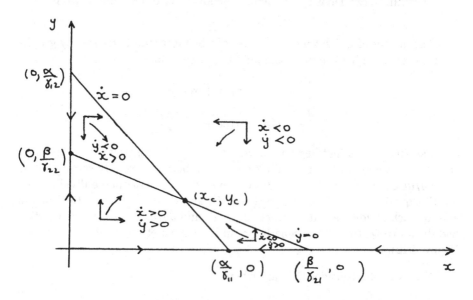

FIGURE 3.13. Division of the first quadrant into 4 distinct regions

Then for $x_c > 0$ we need $\alpha > \gamma_{12}\beta/\gamma_{22}$. As determined previously, this would result in the critical point $(0, \beta/\gamma_{22})$ being unstable. Similarly, $y_c > 0$ requires $\beta > \alpha\gamma_{21}/\gamma_{11}$ and results in $(\alpha/\gamma_{11}, 0)$ being unstable also. The interesting deduction is that the positive critical point only exists if both of the extinction ones are unstable. On the other hand, in the presence of strong competition so that $\Delta < 0$, it only exists if both of the extinction points are stable (see Figure 3.13).

We have already discussed briefly the general criterion for stability of a critical point. It can be shown that the stability criterion is given by the inequalities trace $A < 0$ and det $A > 0$ where

$$A = \begin{pmatrix} a_{11} & a_{12} \\ a_{21} & a_{22} \end{pmatrix}.$$

For our present case, we have seen that

$$A = \begin{pmatrix} -x_c\gamma_{11} & -x_c\gamma_{21} \\ -y_c\gamma_{21} & -y_c\gamma_{22} \end{pmatrix}$$

so that trace $A = -(x_c\gamma_{11} + y_c\gamma_{22})$, which is always less than zero if $x_c > 0$ and $y_c > 0$. Also,

$$\det A = x_c y_c(\gamma_{11}\gamma_{22} - \gamma_{21}\gamma_{12})$$
$$= x_c y_c \Delta$$
$$> 0 \text{ if } \Delta > 0.$$

Hence, as we already suspected, for stability we require $\Delta > 0$. If $\Delta < 0$ we have instability resulting in a saddle point because the roots (eigenvalues λ_1 and λ_2) are real and opposite in sign.

To classify the stable case, we need to determine whether λ is real or complex. Consider the discriminant

$$(\text{trace } A)^2 - 4\det A = (x_c\gamma_{11} + y_c\gamma_{22})^2 - 4x_c y_c(\gamma_{11}\gamma_{22} - \gamma_{12}\gamma_{21})$$
$$= (x_c\gamma_{11} - y_c\gamma_{22})^2 + 4x_c y_c\gamma_{12}\gamma_{21} > 0$$

always if $x_c > 0$ and $y_c > 0$. Hence the roots are real and for the stable case both roots are negative, resulting in a stable node.

Summarizing, for the stable case, we require

(i) $\gamma_{11}\gamma_{22} > \gamma_{12}\gamma_{21}$, i.e., low competition;

(ii) $\alpha > \gamma_{12}\beta/\gamma_{22}$, ensuring impossibility of extinction.

Note that the condition $\beta > \gamma_{21}\alpha/\gamma_{11}$ is made redundant from conditions (i) and (ii).

Summarizing for the unstable case, the conditions are

(i) $\gamma_{11}\gamma_{22} < \gamma_{12}\gamma_{21}$, i.e., high competition;

(ii) $\alpha < \gamma_{12}\beta/\gamma_{22}$;

(iii) $\beta < \gamma_{21}\alpha/\gamma_{11}$.

Conditions (ii) and (iii) ensure that the extinction critical points are both stable, resulting in the death of one of the species. Of course, to find which of the species is dying we need to examine which side of a critical trajectory we start from. When $\Delta < 0$ we are dealing with a saddle point so this critical trajectory is the *stable* axis of the saddle point, in other words, the one on which $\underset{\sim}{x} \to 0$ (see earlier description of a saddle point). To discover where this axis goes to when we get far from the saddle is a computing task but one destination is probably the origin. The problem of one species dying, but of course which one, is of great ecological interest. The answer depends on the initial numbers of each species in contrast to the previous stable case.

We may also determine if the positive critical point is stable or not by graphical means involving minimal calculations. Take the case where (see Figure 3.13)

$$\frac{\gamma_{11}}{\gamma_{21}} > \frac{\alpha}{\beta} > \frac{\gamma_{12}}{\gamma_{22}}.$$

So from the graph, providing there are a positive number of each species to begin with the positive critical point is stable. This is a verification of the result gained using the more analytical method. However, once again, the graph cannot predict the nature of the critical points nor can we sketch the trajectories with any great accuracy.

3.6. Some Predator–Prey Models

(a) A General Predator–Prey Model

Consider the model expressed by a system of differential equations

$$\frac{dx}{dt} = xf(x,y),$$

$$\frac{dy}{dt} = yg(x,y). \tag{3.55}$$

Since

(i) $f(x,y)$ decreases with x due to resource limitation or density dependence of prey,

$$\frac{\partial f}{\partial x} \le 0;$$

(ii) $f(x, y)$ decreases with y due to predation of prey,

$$\frac{\partial f}{\partial y} < 0;$$

(iii) $g(x, y)$ increases with x due to predator–prey interaction

$$\frac{\partial g}{\partial x} < 0;$$

(iv) $g(x, y)$ decreases with y due to possible density dependence of predators,

$$\frac{\partial g}{\partial y} \leq 0,$$

then the nonzero equilibrium position is given by

$$f(x_c, y_c) = 0, \qquad g(x_c, y_c) = 0. \tag{3.56}$$

As discussed in the previous section, the characteristic equation (3.33) in this model becomes

$$\lambda^2 - (a_{11}x_c + a_{22}y_c)\lambda + x_c y_c(a_{11}a_{22} - a_{21}a_{12}) = 0 \tag{3.57}$$

where

$$\begin{aligned} a_{11} &= f_x(x_c, y_c), & a_{12} &= f_y(x_c, y_c), \\ a_{21} &= g_x(x_c, y_c), & a_{22} &= g_y(x_c, y_c). \end{aligned} \tag{3.58}$$

In view of the explanation given above about the signs of the derivatives, it is clear that equation (3.57) can be written as

$$\lambda^2 + A\lambda + B = 0 \qquad (A \geq 0, B > 0). \tag{3.59}$$

This shows that the equilibrium is neutral if $A = 0$ or if $a_{11} = 0$, $a_{22} = 0$, i.e., if there is no density dependence, and the equilibrium is stable if a_{12} or a_{22} or both are negative, i.e., if there is density dependence.

(b) May's Predator–Prey Models

R.M. May has proposed a more general class of models for two interacting populations which includes the classical Lotka–Volterra model in a special case. Although the Lotka–Volterra differential equations (3.1) involving the prey population $x(t)$ and the predator population $y(t)$ have long been used by ecologists, relatively few realistic models can be obtained using the various terms in the equations. As proposed by May (1981), a more realistic predator–prey model can be expressed by the equations

$$\begin{aligned} \frac{dx}{dt} &= rx\left(1 - \frac{x}{K}\right) - yf(x, y), \\ \frac{dy}{dt} &= yg(x, y). \end{aligned} \tag{3.60}$$

If in particular $f(x,y) = \alpha x$ and $g(x,y) = b - \beta x - \gamma y$ we get the classical Lotka–Volterra equations (3.1). Some other forms for the functions $f(x,y)$ and $g(x,y)$ have been discussed by May (1981). It is been shown that essentially all such predator–prey models have either a stable point or a stable limit cycle.

Let us consider one particular model, constructed from equations (3.60) with the forms for $f(x,y)$ and $g(x,y)$ respectively as

$$f(x,y) = \frac{\kappa x}{x + D},$$

$$g(x,y) = s\left(1 - \frac{y}{\gamma x}\right).$$

(3.61)

We thus obtain

$$\frac{dx}{dt} = rx\left(1 - \frac{x}{K}\right) - \frac{\kappa xy}{(x + D)},$$

$$\frac{dy}{dt} = sy\left(1 - \frac{y}{\gamma x}\right).$$

(3.62)

The equilibrium position for model (3.62) is given by

$$x_c = \frac{D(1 - \alpha - \beta + R)}{2\beta},$$

$$y_c = \gamma x_c$$

(3.63)

where

$$\alpha = \kappa\gamma/r, \qquad \beta = D/K, \qquad R = \left[(1 - \alpha - \beta)^2 + 4\beta\right]^{\frac{1}{2}}.$$

(3.64)

It can be shown that the characteristic equation for (3.62) is

$$\lambda^2 + \left\{s - rx_c\left[\frac{\alpha x_c}{(x_c + D)^2} - \frac{1}{\kappa}\right]\right\}\lambda + rsx_c\left(\frac{1}{\kappa} + \frac{\alpha D}{x_c + D)^2}\right) = 0.$$

(3.65)

The equilibrium is therefore stable if

$$\frac{s}{r} > \frac{\alpha - R}{1 + \alpha + \beta + R}.$$

For the choice of parameters

$$\frac{r}{s} = 6, \qquad \alpha = 1, \qquad \beta = 0.1,$$

May has shown that the system exhibits stable cycles. Other parameter choices could also give stable equilibrium point solutions.

(c) Difference Equation Models

Consider the logistic equation (3.5) in the form

$$\frac{dN}{dt} = aN - bN^2 \tag{3.66}$$

where $N(t)$ is the population size at time t. The solution to the logistic equation as in (3.7) is given by

$$N(t) = \frac{Lae^{at}}{1 + Lbe^{at}} = \frac{a/b}{1 + e^{-at}/Lb} \tag{3.67}$$

where L is a constant. If we put $Lb = e^{-at_0}$, then

$$N(t) = \frac{a/b}{1 + e^{-a(t-t_0)}}. \tag{3.68}$$

We can now express (3.68) as a difference equation, that is one that gives population size at time $t+1$ in terms of its size at time t. Equation (3.68) can be rewritten as

$$N(t) = \frac{\kappa}{1 + J\lambda^{-t}} \tag{3.69}$$

where $\kappa = a/b$ is the population's saturation level, $J = e^{at_0}$ and $\lambda = e^a$. Clearly,

$$N(t+1) = \frac{\kappa}{1 + J\lambda^{-1(t+1)}} = \frac{\kappa\lambda}{\lambda + J\lambda^{-t}}. \tag{3.70}$$

We can thus derive the difference equation

$$N(t+1) = \frac{\lambda N(t)}{1 + \alpha N(t)} \tag{3.71}$$

where $\alpha = (\lambda - 1)/\kappa$.

This technique contains the basic difference equation formulation

$$N(t+1) = f(N(t)) \tag{3.72}$$

where f is an arbitrary function. Equation (3.72) can exhibit some rather remarkable behaviour, a fact of considerable importance because it has seen, and will continue to see, a great deal of practical significance. In many ways the discrete form of the model is easier to work with than the continuous form.

We now extend the discussion by considering populations of two species living together and competing with each other for the same limiting resource i

$(i = 1, 2)$. The differential equations governing them are as in (3.1)

$$\frac{dN_1}{dt} = a_1 N_1 - b_{11} N_1^2 - b_{12} N_1 N_2,$$
$$\frac{dN_2}{dt} = a_2 N_2 - b_{21} N_1 N_2 - b_{22} N_2^2. \tag{3.73}$$

The above simultaneous differential equations cannot, in general, be solved explicitly. Hence, it is necessary to find difference equations that permit prediction of the sizes of the two populations at time $t + 1$, given their sizes at time t. We thus require expressions corresponding to (3.72) of the form

$$N_i(t + 1) = f_i\left(N_1(t), N_2(t)\right) \quad \text{for} \quad i = 1, 2.$$

Thus, knowing $N_1(0)$ and $N_2(0)$, we can calculate successively $N_1(1)$, $N_2(1)$, $N_1(2)$, $N_2(2)$, and so on. When the conditions are such that there is unstable equilibrium we may find out by this method which of the two competing species will be the winner for given values of $N_1(0)$ and $N_2(0)$.

Furthermore, the differential equations (3.73) may be replaced by the difference equations as shown before in the logistic model

$$N_1(t + 1) = \frac{\lambda_1 N_1(t)}{1 + \alpha_1 N_1(t) + \beta_1 N_2(t)},$$
$$N_2(t + 1) = \frac{\lambda_2 N_2(t)}{1 + \alpha_2 N_2(t) + \beta_2 N_2(t)}. \tag{3.74}$$

It can be shown that (3.74) is equivalent to (3.73) provided both α_i and β_i are proportional to $\lambda_i - 1$. In fact, the relationship between the parameters of the difference equations and the differential equations is given by

$$\lambda_1 = e^{a_1}, \qquad \alpha_1 = b_{11}(e^{a_1} - 1)/a_1, \qquad \beta_1 = b_{12}(e^{a_1} - 1)/a_1$$

and so on.

Thus, if the parameters in (3.73) are given, those in (3.74) may easily be obtained. Hence, for any preassigned combination of the initial population $[N_1(0), N_2(0)]$, the sequence $[N_1(t), N_2(t)]$ for $t = 1, 2, \ldots$ may be calculated and the trajectory of the entire population plotted. Using different starting points leads to a whole family of such trajectories.

Problems

1. Solve

$$\frac{dX}{dt} = -X - 6Y,$$

$$\frac{dY}{dt} = -X - 2Y,$$

classifying the critical point, and sketching trajectories. Suppose the competing-species model has such a critical point, with $X = x - x_c$, $Y = y - y_c$, and $x_c, y_c > 0$. What can you say about the ultimate fate of x and y?

2. (i) Solve the set of simultaneous ordinary differential equations

$$\frac{dX}{dt} = -X - 3Y,$$

$$\frac{dY}{dt} = -2X - 2Y.$$

(ii) What is meant by a *trajectory* of the above system? Describe the behaviour of the trajectories of the system as $t \to +\infty$ and $t \to -\infty$.

(iii) Suppose $x(t)$, $y(t)$ are the numbers of two competing species which have an equilibrium point (x_c, y_c) where x_c, y_c are both positive. If $X = x - x_c$, $Y = y - y_c$ satisfy the above equation, roughly sketch the trajectories in the first quadrant of the (x, y) plane. Discuss, *in detail*, the ultimate fate of x and y.

3. Consider the set of non-linear differential equations

$$\frac{dx}{dt} = x - xy,$$

$$\frac{dy}{dt} = -y + xy.$$

(i) What is meant by an *equilibrium point* of such a system? Show that the origin and the point $(1, 1)$ are equilibrium points of the above system.

(ii) Show that $(0, 0)$ is a saddle point and $(1, 1)$ is a centre of the above system.

(iii) Roughly sketch the trajectories in the first quadrant of the (x, y) plane and discuss the ecological implications if $x(t)$ and $y(t)$ represent the numbers of two competing species.

4. Consider two species whose survival depends upon their mutual cooperation. A simple model for this situation is given by the autonomous system:

$$\frac{dx}{dt} = -ax + bxy,$$

$$\frac{dy}{dt} = -cy + dxy.$$

   (i) Interpret the constants a, b, c and d and obtain the equilibrium levels.

   (ii) Sketch the trajectory directions in the phase plane.

   (iii) Find an analytical solution. Do you believe the model is realistic?

5. The differential equations

$$\frac{dx}{dt} = x(1 - x - y),$$

$$\frac{dy}{dt} = y(2 - x - y)$$

describe populations (x, y) of two competing species. Sketch trajectories and prove that no matter what are the initial (positive) populations of x and y, in the limit as $t \to \infty$, x becomes extinct.

6. The equations describing competitive interaction between two populations are

$$\frac{dx}{dt} = x \left\{ \alpha - (\gamma_{11}x + \gamma_{12}y) \right\},$$

$$\frac{dy}{dt} = y \left\{ \beta - (\gamma_{21}x + \gamma_{22}y) \right\}.$$

If $\gamma_{21} < \beta\gamma_{11}/\alpha$ and $\gamma_{12} < \alpha\gamma_{22}/\beta$, show that there is a nontrivial equilibrium point, and prove that it is stable. Discuss the ecological significance of this result. Discuss also the case when $\gamma_{21} > \beta\gamma_{11}/\alpha$ and $\gamma_{12} > \alpha\gamma_{22}/\beta$.

7. Write the Lotka–Volterra type equations for a system of two predators and two prey in which the prey do not compete with each other but both predator species eat both prey species. The predators should not have a self-damping term, the prey should have a self-damping term. Obtain the equilibrium values for the prey species.

4

Biogeography: Mathematical Analysis of Wildlife Reserves

4.1. Basic Concepts

The study of biogeography is predominantly concerned with conserving or determining the maximum number of species possible on an island. This is achieved through setting up mathematical models, both static and dynamic, whose parameters have been experimentally evaluated and statistically tested. These parameters will usually be assumed to be constant so as to enable easier deductions from the models from some of the more important variables. These include the number of species, area of the island and, for the dynamic models, immigration and extinction rates. Islands are simpler than continents and by examining them in clusters we may gain insight into the more complex topic of continental biogeography. Also, evolutionary hypotheses are more easily tested on islands as evolution is more likely to take place more rapidly in small isolated populations.

Because of the sizeable number of islands and their varied shapes and sizes they lend themselves to statistical survey and hence may facilitate the use or building of mathematical models. The number of species on different islands may vary considerably and this is due to a number of influences.

There may be problems of access: oceans, for example, may create barriers for nonflying organisms. Larger islands will usually support more species as they will contain a larger variety of habitats than the smaller islands.

An island isolated from the mainland is less likely to have a large variety of life than a closer island and the richness of the mainland source will also affect this variety.

The number of species on an island is also less than that for a similar area on the mainland due to natural hazards such as volcanic eruptions where vacating the area may not be possible, as it is on the mainland. Also, re-invasion for chance extinction may not be as simple on an island.

4.2. Biogeographical Variables for Islands

Our main concern is the number of species found on an island. The main factors affecting this are as follows:

- area and topography
- accessibility of colonists from sources and richness of that source
- rates of colonization by new species and rates of extinction by existing species

We shall denote these symbolically as follows:

$S =$ number of species on an island
$A =$ area of island
$I =$ rate of immigration of new species
$E =$ rate of extinction of existing species

These four quantities will form the basis of two basic models, one static and the other dynamic.

4.3. The Species–Area Relationship

A static approach relating the number of species in an island biota to its area is given by the species–area relationship (see Figure 4.1)

$$S = cA^z \tag{4.1}$$

where

$S =$ number of species of a given taxon (a particular group) found on an island

$A =$ area of the island

$c =$ a parameter depending on taxon and biogeographic regions

$z =$ parameter with little variation among taxons.

For virtual islands, the smaller value of z is due to the flood of transient species, that is, species which maintain themselves ecologically nearby.

Thus, virtual islands carry more species than an island of the same size. This is due to the fact that virtual island species need not be well-adapted to that habitat to survive there, provided that they persist in places close at hand.

So we have, in effect, more species per unit area and hence a smaller z-value.

4.4. Equilibrium Theory

We may equate the rates of immigration and extinction to achieve dynamic equilibrium on an island. We shall denote by S^* the equilibrium number of species.

Assumptions required to develop this theory are the following:

1. The immigration rate, I, decreases with increasing S, the number of species present on the island.

2. For fixed S, I decreases with increasing distance D from the mainland.

3. The extinction rate, E, increases as S increases.

4. For fixed S, E depends on A, the area of the island.

A smaller island will have a smaller population of each species it can support and hence the risk of extinction for any one of them increases.

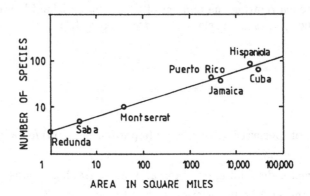

FIGURE 4.1. Species–area curve for West Indian herpetofauna (amphibians and reptiles)

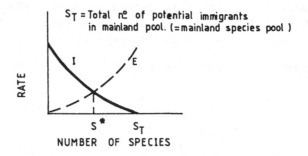

FIGURE 4.2. Extinction rate (E) and immigration rate (I) against the number of species

The required equilibrium situation is best portrayed diagrammatically (see Figure 4.2). Hence when immigration and extinction rates are equal the equilibrium number of species for the island is given by S^*.

If immigrations were to increase above the number of extinctions, S would increase above S^* and in consequence, I would tend to decrease and E would increase (and vice versa) forcing S back to equilibrium. Hence we have a stable equilibrium situation.

The following diagrams, Figures 4.3 (a)–(b), summarize the effects of location and size of an island on S.

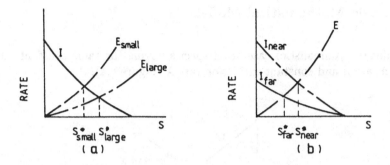

FIGURE 4.3. Effects of location and size of an island

The usual range of z for oceanic islands is from 0.20 to 0.35 whilst for virtual islands such as hilltops and valleys z varies from 0.12 to 0.17. This is indicated in the following Table 4.1.

TABLE 4.1. Values of z for different islands

Fauna or Flora	Island Group	z
Carabid beetles	West Indies	0.34
Ponerine ants	Melanesia	0.30
Amphibians and reptiles	West Indies	0.301
Breeding land and freshwater birds	West Indies	0.237
Breeding land and freshwater birds	East Indies	0.280
Breeding land and freshwater birds	East-Central Pacific	0.303
Land vertebrates	Islands of Lake Michigan	0.239
Land plants	Galapagos Islands	0.325

The data are usually fitted logarithmically, i.e.

$$\log S = \log c + z \log A. \qquad (4.2)$$

Hence a larger island at the same distance from the mainland has a lower extinction curve and so has a larger S^*. Also, an island of the same size but farther from the mainland has a lower immigration curve and so has a smaller S^*.

However, it should be pointed out that this is a rather simplified approach and fails to account for various factors such as evolution, the behaviour of I for small S (slow initial immigration) and the possibility of E and I being dependent. This description also does not account for the topography of the island and treats all species as equal.

4.5. Basic Mathematical Models

Consider the (unrealistic) case of all species having the same constant immigration rate μ and constant extinction rate λ (Figure 4.4).

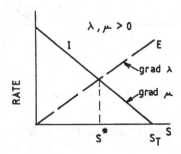

FIGURE 4.4. The immigration and the extinction function as linear approximation

Although simple, this model may provide a sensible linear approximation in the neighbourhood of equilibrium to a more general nonlinear model.

The model is formulated as follows:

Net extinction rate

$$E = \lambda S. \tag{4.3}$$

Net immigration rate

$$I = \mu(S_T - S) \tag{4.4}$$

where

S_T = total number of potential immigrants in the mainland pool

$S_T - S$ = number of candidates for immigration to island on which S species are already present.

At equilibrium, we have $E = I$, determining S^* as

$$S^* = \left(\frac{\mu}{\lambda + \mu}\right) S_T. \tag{4.5}$$

From equation (4.5) we see that S^* increases when the immigration gradient μ increases and S^* decreases when the extinction gradient λ increases, as expected.

If the island is not in equilibrium, then the change in the number of species per unit time is the difference between immigration and extinction rates, i.e., we have the nonequilibrium adjustment equation

$$\frac{dS}{dt} = I - E. \tag{4.6}$$

Substituting (4.3) and (4.4) into (4.6) gives

$$\frac{dS}{dt} = \mu S_T - (\lambda + \mu)S. \tag{4.7}$$

Integrating (4.7) with respect to t, and letting $S(0)$ be the number of species initially on the island, we obtain

$$S(t) - S^* = (S(0) - S^*) e^{-(\lambda + \mu)t}. \tag{4.8}$$

This derived expression describes the rate at which the system approaches the equilibrium value S^*.

This model can be applied to a definition by Diamond (1975) of the relaxation time t_r of an island biota. This is merely the time required for the deviation from equilibrium to decrease to e^{-1}, i.e., 36.8% of its initial magnitude. Thus, we have

$$\frac{S(t) - S^*}{S(0) - S^*} = e^{-1} = e^{-(\lambda + \mu)t_r} \tag{4.9}$$

which yields

$$t_r = \frac{1}{\lambda + \mu}. \tag{4.10}$$

To make a 90% recovery (for an island in disequilibrium), we have

$$0.10 = e^{-(\lambda + \mu)t_{0.90}} \tag{4.11}$$

which yields

$$t_{0.90} = \frac{2.3026}{\lambda + \mu} = 2.3026 t_r, \tag{4.12}$$

that is, 90% recovery occurs during the third relaxation period of the island.

There are a few basic drawbacks with this model. We have assumed that λ and μ are constants for all species, which gives linear immigration and extinction rates. In fact, extinction rates increase faster than linearly with S and net immigration rates fall faster than linearly with S. This is predominantly due to the fact that late arrivals tend to be poor dispersers, which slows immigration; and the extinction rate is increased due to competitive exclusion, that is, a species may be far less competitive than another in its own habitat.

In general, due to competition and so forth, μ decreases and λ increases with S, extinction dominating newer immigrants. An extreme view of this situation was put forward by Lack in 1973 and is described briefly here.

Suppose a number of species S^* are predominantly resident on an island and that other immigrants will fail to breed, that is, there are a particular number, S^*, of species that are ecologically appropriate to the island. This is formulated diagramatically as follows:

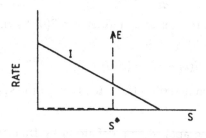

FIGURE 4.5. Immigration and extinction curves for ecologically appropriate species

Extinction rates for S^* species ecologically appropriate to the island are effectively zero, and for any other species extinction is infinite.

The real situation for immigration and extinction curves usually lies between the above two cases, explained graphically in Figures 4.4 and 4.5. As more and more species are packed into an island the island approaches ecological saturation and clearly the extinction curve steepens rapidly. We can therefore model the real situation more accurately by use of the modified extinction function

$$E(S) = \epsilon \left(\frac{S}{S^*} \right)^n \tag{4.13}$$

where n is a parameter indicating the steepness of the extinction curve and ϵ acts much in the same manner as λ.

For birds on an (oceanic) island n varies from 3 to 4.

This is a generalization of our other models. For $n = 1$, equation (4.13) reduces to

$$E(S) = \left(\frac{\epsilon}{S^*} \right) S, \tag{4.14}$$

which is analogous to our basic linear model, given by equation (4.3).

For $n \to \infty$ we obtain our other extreme

$$E(S) \to \begin{cases} 0, & S < S^*; \\ \infty, & S > S^*. \end{cases} \tag{4.15}$$

Analogous expressions to these may be used to parameterize the nonlinear behaviour of most curves for net immigration $I(S)$, e.g., for birds on (oceanic) islands $n \sim 6$. For comparison, we now look at a rate diagram provided by Gilpin and Diamond in 1976.

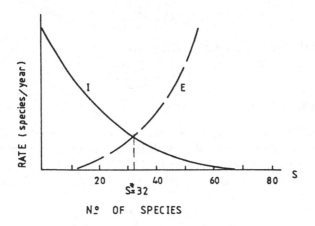

FIGURE 4.6. Lowland bird species on Three Sisters, one of the smaller Solomon Islands

The extinction curve here is of the form

$$E(S) = \epsilon \left(\frac{S}{S^*} \right)^{2.75} \quad \text{where } S^* = 32. \tag{4.16}$$

MacArthur and Wilson (1967) have shown that the extreme case shown in Figure 4.5 gives a better approximation to the real situation than the simple linear model discussed at the beginning of this section.

In 1976, Gilpin and Diamond worked on numerous versions of the equilibrium model by choosing certain functional forms for E and I. In the particular model discussed here it is assumed that E is a function of S and A in separable form, i.e.,

$$E(S, A) = g(S)e(A) \tag{4.17}$$

and I depends on S and D (distance from mainland source) in much the same manner as above, that is,

$$I(S, D) = h(S)i(D). \tag{4.18}$$

Due to the concavity of the curve for E vs. S (in the rate diagram), it is assumed that $g(S)$ increases with S and also that the rate of $g(S)$ increases with S, i.e., $g'(S) > 0$, $g''(S) > 0$. This suggests the form

$$g(S) = E_0 S^n, \quad n > 1 \tag{4.19}$$

where E_0 is a constant of proportionality.

Next we assume that E is inversely proportional to A and so, without loss of generality, we put

$$e(A) = \frac{1}{A} \tag{4.20}$$

which yields the two-parameter (n, E_0) model

$$E(S, A) = \frac{E_0 S^n}{A} \tag{4.21}$$

for the extinction curve.

Now consider $h(S)$ and $i(D)$. As these models were tested on bird species, we can assume that the curve of I vs. S is concave (as in the rate diagram), which implies that $h'(S) < 0$, $h''(S) > 0$. Also, $h(S)$ is a decreasing function of S/S_T, the fraction of the mainland species already on the island. This suggests the form

$$h(S) = I_0 \left(1 - \frac{S}{S_T} \right)^m, \quad m > 1 \tag{4.22}$$

where I_0 is a constant of proportionality.

For $i(D)$, Gilpin and Diamond took the form

$$i(D) = \exp \left[-\frac{\sqrt{D}}{D_0} \right] \tag{4.23}$$

as it fitted the data well and had only one parameter, D_0. This gives the three-parameter (m, I_0, D_0) model

$$I(S, D) = I_0 \left(1 - \frac{S}{S_T} \right)^m \exp \left[-\frac{\sqrt{D}}{D_0} \right] \tag{4.24}$$

for the immigration curve.

To solve for the equilibrium value S^* we put $E(S, A) = I(S, D)$, which yields

$$\frac{E_0 S^n}{A} = I_0 \left(1 - \frac{S}{S_T} \right)^m \exp \left[-\frac{\sqrt{D}}{D_0} \right]. \tag{4.25}$$

Clearly, without loss of generality, we may put $I_0 = 1$ and $E_0 = R$, say. Also, through empirical testing Gilpin and Diamond gave a model with $m = 2n$ which produced a good fit. This is supported by evidence showing that immigration curves are far more concave than extinction curves. Hence we now have the simplified three-parameter model

$$E(S, A) = \frac{RS^n}{A} \qquad (4.26)$$

and

$$I(S, D) = \left(1 - \frac{S}{S_T}\right)^{2n} \exp\left[-\frac{\sqrt{D}}{D_0}\right] \qquad (4.27)$$

with parameters R, n and D_0.

This model was tested by Gilpin and Diamond on lowland birds of the Solomon Islands, where it was known that $S_T = 106$. The rate diagram is given in Figure 4.7.

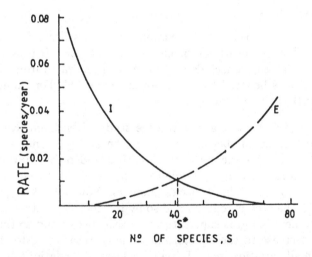

FIGURE 4.7. Three-parameter model with $R = 1.49 \times 10^{-5}$, $n = 2.37$, $D_0 = 2.11$

The area of the island was considered to be $A = 10$ square miles and its distance from the mainland $D = 25$ miles. The intersection is at $S^* = 40.4$. This model fitted the data well, explaining 98.6% of the observed variance in S. This implies that, at least for the data used, almost all variation in S is due to variation in A and D. The islands and fauna were, however, chosen deliberately to be almost wholly determined by A and D.

Models of the type discussed in this chapter are predominantly dynamic, due to the relationship

$$\frac{dS}{dt} = I - E,$$

and we can examine the different approaches of the number of species toward equilibrium. There are three possible cases for our formulation: (i) approach from above $S(0) > S^*$; (ii) from below $S(0) < S^*$; or (iii) staying within a neighbourhood $S(0) \approx S^*$. All of these cases occur naturally.

Consider first the case of the number of species increasing towards the equilibrium value. Natural catastrophes such as tidal waves or volcanic eruptions are usually the causes of this situation. In this case the immigration term dominates, at least initially, in the restabilizing of the island. For the case of the number of species staying in the neighbourhood of the equilibrium value, the actual number of species shows small, balanced fluctuations due to changes in identities of species. That is to say that new species are constantly immigrating and other species becoming extinct. This may possibly be due to competitive exclusion.

The final case is that of the number of species decreasing towards the equilibrium value. A natural occurrence of this situation is provided by land-bridge islands. These are islands that have been separated from continents or from larger islands believed to be previously connected. For example, Britain in Europe or Borneo and Japan in Asia.

Terborgh (1974) made a quantitative study of birds on five land-bridge islands. The current number of bird species on all five islands was unknown. The number of bird species present on each island before it was cut off by rising sea levels 10 000 years ago (due to an Ice Age) was taken as an estimate of the neighbouring mainland species numbers. Now, after being cut off from the mainland, the area is reduced and so by the species–area relationship, i.e., $S = cA^z$, the new island is supersaturated with species and so the number of species will decrease to the equilibrium value corresponding to the new area (assuming small variation in z). Initially at least, the relationship

$$\frac{dS}{dt} = I - E$$

will be dominated by the E term so we can assume initially that

$$\frac{dS}{dt} = -E. \tag{4.28}$$

Terborgh chose $n = 2$ for steepness of extinction in equation (4.13), that is

$$E(S) = kS^2, \quad k > 0 \tag{4.29}$$

and hence using his data in the relationship

$$\frac{dS}{dt} = -kS^2 \tag{4.30}$$

he determined the extinction parameter as a function of island area. The extinction parameter k thus determined is shown in Figure 4.8 with a straight line of best fit.

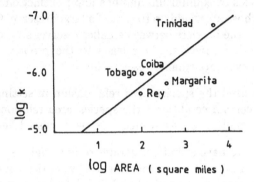

FIGURE 4.8. The extinction parameter k as a function of island area for birds on land-bridge islands in the West Indies

An application of these calculations correctly predicted extinction rates observed on Barro Colorado Island which has been carefully protected as a wildlife reserve since 1923.

4.6. Design of Wildlife Reserves

The more artificial case of the number of species decreasing towards equilibrium, as discussed in the previous section, is given by the creation of reserves. That is, a fraction of a habitat is set aside as a reserve and the rest is destroyed, leading to supersaturation of species on the island.

Clearly, our main interest is to optimally design the reserves in order to maximize the equilibrium number of species there. The survival of individual species depends on conservation strategies. If we suppose that each species has equal probability of survival, then a large number of small reserves would be satisfactory. The reason for this is that each reserve would lose most of its species before reaching equilibrium but with enough reserves any given species should survive in at least one reserve. This seems ideally simple and in fact does not hold true. The basic flaw lies in the fact that different species have different area requirements for survival, arising from very different rates of immigration and extinction. For example, a species incapable of dispersion from one reserve to another faces extinction whereas a species capable of dispersal may persist due to a balance between local immigration and extinction (if recolonization rates are high enough).

It is found that in general, a large reserve is better than a smaller reserve as it holds more species at equilibrium and has lower extinction rates. If we take smaller reserves whose areas add up to that of a larger reserve, the larger reserve is usually better as the barriers between smaller reserves stop dispersal of some species into the other reserves. This leads to the problem to be discussed further – when, if ever, are smaller reserves better?

Higgs (1981) used the species–area relationship in setting up a model to show that under certain conditions, the species–area relationship may favour the setting up of smaller reserves, which is to be discussed here.

Suppose that we have a limited amount of financial resources, so that we can conserve only A units of area. Consider the two options of either taking one large area A or two smaller areas of pA and $(1-p)A$ where $0 < p < 1$. Assuming the species–area relationship on natural reserves, the number of species, S_1, appearing on the one large reserve is given by

$$S_1 = cA^z. \tag{4.31}$$

The total number of species, S_2, on the two smaller reserves is given by

$$S_2 = c(pA)^z + c\{(1 - p)A\}^z - v \tag{4.32}$$

where v is the number of species common to the two smaller areas (often called the "overlap").

If the same number of species are preserved in either situation then $S_1 = S_2$ and hence

$$v = \{p^z + (1 - p)^z - 1\} cA^z. \tag{4.33}$$

If v is smaller than the right-hand side value, more species will be preserved in the two smaller reserves and vice versa. For convenience, define the proportional overlap of the two smaller reserves as

$$P_v = \frac{v}{S_2}. \tag{4.34}$$

By substituting (4.32) and (4.33) into (4.34) we obtain the form

$$P_v = p^z + (1 - p)^z - 1. \tag{4.35}$$

The results are shown in Figure 4.9.

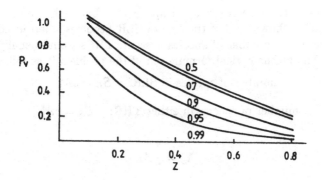

FIGURE 4.9. Graph of isoclines (values of p) for P_v vs. z

When the actual value P_v is less than that calculated from equation (4.35) then two smaller reserves are favoured. Higgs has shown, for example, that for z about 0.3 with the two areas not differing by more than about half an order of magnitude in size, there will be more species on the two smaller reserves if the actual proportional overlap is less than approximately 0.6.

It is found that the favourable case for two smaller reserves applies to three quarry nature reserves shown in Table 4.2.

TABLE 4.2. A comparison of three quarry nature reserves in the chalk (Cretaceous) of the Yorkshire Wolds (KCP, Kiplingcotes Chalk Pit; RBQ, Rifle Butts Quarry; WQ, Wharram Quarry)

Reserves compared	p	P_v estimated field lists	P_v calculated from equation (4.35)
RBQ, WQ	0.94	0.40	0.44
WQ, KCP	0.59	0.43	0.65
RBQ, KCP	0.92	0.45	0.48

A value of $z = 0.276$ has been used. If P_v from field lists is less than the calculated value of P_v, then two smaller reserves are favoured.

Although the benefits shown towards smaller reserves in the table are small, there is no evidence, for housing species on reserves, to suggest that one large quarry is the optimal strategy.

In another study, Higgs (1981) considered the case of two equal-sized half reserves (THR) in place of a single large reserve (SLR) of the same total area using a simpler, clever mathematical approach. Using the species–area approach it can be shown that if the size of SLR becomes small in comparison with the size of the number of species present, THRs would usually be more favourable. The mathematical formulation of the problem is as follows:

$$\text{number of species in SLR:} \quad S_L = cA_L^z \qquad (4.36)$$

$$\text{number of species in either THR:} \quad S_S = cA_S^z. \qquad (4.37)$$

But

$$A_S = \frac{1}{2}A_L \qquad (4.38)$$

yields

$$S_S = 2^{-z}S_L. \qquad (4.39)$$

Let v be the number of species shared by the THRs and P_v be the proportion of species within THRs that are shared. Then

$$P_v = \frac{\text{number of shared species}}{\text{total number of species in both THRs}} = \frac{v}{2S_S - v}. \qquad (4.40)$$

Substitute for S_S from equation (4.39) into equation (4.40) and rearrange to obtain

$$v = \frac{P_v 2^{1-z}S_L}{1 + P_v}. \qquad (4.41)$$

If S_T denotes the total number of species in THRs, then

$$S_T = 2S_S - v. \qquad (4.42)$$

Substituting from (4.39) yields

$$S_T = 2^{1-z}S_L - v. \qquad (4.43)$$

Now define the ratio

$$R = \frac{S_T}{S_L}. \qquad (4.44)$$

Thus

$$R = \frac{2^{1-z}S_L - v}{S_L} \qquad (4.45)$$

and using equation (4.41) we obtain the form

$$R = \frac{2^{1-z}}{1 + P_v}. \qquad (4.46)$$

A plot of isoclines (for R) with P_v against z is shown in Figure 4.10.

We can see from the above diagram that the lower the proportion of species in common, the steeper the slope of the species–area curve must be to favour an SLR.

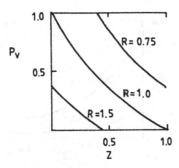

FIGURE 4.10. Slope of species–area curve for the THR case

The case $R = 1$ in this formulation shows no preference for either case.

We can further look at the effect of the species-pool size for the case $R = 1$ (the species-pool being the number of species able to immigrate to the island, whether they could persist there or not).

Usually, values of P_v will depend on the size of the reserve. As reserve size decreases, fewer species in the species-pool will be able to persist there, which gives lower values for P_v.

Assuming the simple model where species are distributed randomly in the THRs, it can be shown that

$$P_v = \frac{2^{-2z}T}{2 - 2^{-2z}T} \tag{4.47}$$

where T is the proportion of the species-pool contained in the SLR, i.e.,

$$T = \frac{S_L}{N_P} \tag{4.48}$$

where N_P is the total number of species in the pool. The graphs in Figure 4.11 show the effect of the size of the species-pool on $R = 1$ for isoclines $T = 1, 0.5$.

By halving T, i.e. halving S_L, P_v is reduced, as in the diagram, quite dramatically – especially for small values of z.

The smaller the size of the reserves, the lower the values of P_v, so THRs will be favoured more strongly as the total size of the reserve decreases.

The approaches made so far in this section use the species–area relationship and are weak in the sense that we have assumed throughout that the slope of the species–area curve, z, is constant. Also, in the latter discussion, the proportion

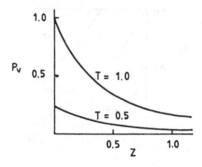

FIGURE 4.11. Effect on the size of the species–area curve

of species in common, P_v, varies from situation to situation and hence may not be entirely accurate for R close to 1 for example.

Of course there are many more considerations to be taken into account and a brief outline of some of the more common factors involved follows.

Quite often we can change from a large reserve to smaller reserves but fragmentation proves to be irreversible and can be quite costly.

Natural catastrophes show no clearcut advantage to any particular strategy. A catastrophe may affect an entire large reserve whereas smaller, separated reserves may not all be affected. However, a large reserve is less likely to be destroyed than a smaller one and recolonization in a large reserve is easier than recolonization of a smaller reserve from another.

A large reserve may reduce the detrimental effects of pollution whereas smaller reserves are more open to pollution from outside. Smaller reserves have more "edge", which may suit certain species that live on edges of reserves, but this requires biological observation to determine an appropriate strategy. It is clear that the relative importance of each factor in the choice of a strategy can only be determined by careful consideration of the biological and managerial consequences in each particular situation. There are also a range of other relevant criteria apart from size which must be considered.

Having examined relationships between species, area and extinction rates and the like, we can finally look at designing the reserves to minimize extinction rates. The designs which follow are such that extinction rates will be lower on average for the reserve design on the left than for that on the right. This is explained in Figure 4.12.

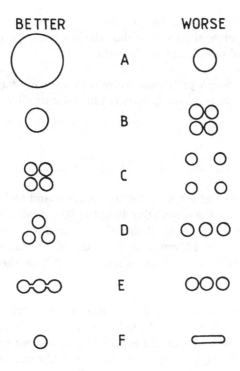

FIGURE 4.12. General design principles

A. A large reserve is better than a smaller reserve as it holds more species at equilibrium and has lower extinction rates.

B. Given a certain total area available for a reserve, the reserve should be divided into as few disjoint pieces as possible, mainly to accord with the underlying principles of A.

C. If the available area must be broken into several smaller reserves, then they should be as close together as possible to increase immigration rates between the reserves.

D. If there are several smaller reserves they should be grouped equidistantly from each other to enable the highest possible immigration rates between them. Placed in a line, the reserves on either end will have smaller immigration levels between them.

E. Several smaller reserves could benefit from being connected with strips of protected habitat enabling more dispersion between reserves.

F. Any given reserve should be as close to circular as possible since this minimizes dispersal distances within the reserve. An elongated reserve may suffer local extinctions at the ends.

These general design principles are subject to many factors. The developement of techniques is currently required in order to fill the designated conservation purpose.

Problems

1. Suppose that we begin with a totally vacant island and that each day 40 species arrive. On the second day 10 of the 40 arrivals were already on the island. On the third day 15 of the 40 were already on the island. On the fourth day 20 of the 40 were already on the island. Graph the number of new arrivals against the number of species already on the island, assuming that all arrivals remain.

2. The island in Problem 1 was 3 miles from the mainland source. Suppose we consider a second island, 15 miles from the mainland source. Because this island is farther from the source of arrivals, we must expect fewer arrivals per day. Suppose that only one-fourth the number arriving on the first island arrive on the second island. If the fraction of arrivals already present is the same as in Problem 1, graph the number of new arrivals against the number of species already on the island.

3. Suppose that an island that originally contained 40 species is artificially isolated and no further immigration from the mainland is possible. If we keep track of the island for four days, we will find that on the first day 16 species will become extinct; on the second day, 10 species; on the third day, 6 species; on the fourth day, 4 species; and on the fifth day 1 species will become extinct. Graph the number of species that will become extinct per day against the number of species on the island.

4. Suppose we have a series of five islands, all the same size. Suppose we isolate these islands and observe them for a year. Suppose that the islands contained 40, 30, 20, 15 and 10 species at the beginning of the year and 35, 27, 18, 13 and 9, respectively, at the end of the year. Graph the extinction rate against the number of species.

5. Compute the equilibrium number of species graphically for the immigration function of Problem 1 and the extinction function of Problem 4 above.

6. Assume that the immigration function is expressed by

$$\frac{dS}{dt} = I_0 - \frac{I_0}{T}S$$

and the extinction function by

$$\frac{dS}{dt} = bS,$$

where I_0 is the rate of immigration to an empty island, T is the total number of species in the mainland, S is the number of species on the island and b is the slope of the extinction function. Show that the equilibrium number of species is given by

$$\hat{S} = \frac{TI_0}{bT + I_0}.$$

Explain this result by graphs.

Suppose that I_0 is linearly related to the distance from the mainland. Show that in this case

$$\hat{S} = \frac{T(D^* - D)}{D^*(b + 1) - D},$$

where D is the distance from the mainland and D^* is the critical distance beyond which no species can immigrate.

5

Pharmacokinetics: Drug Distribution in Pharamacology

5.1. Some Simple Drug Distribution Problems

Consider a simple problem in pharmacology concerned with the dose – response relationship of a drug which has a simple manner of disappearance from the body. We will first discuss a model where the rate of decrease of concentration of the drug is directly proportional to its amount present in the system. Let $C(t)$ denote the concentration of the drug in the system at time t and C_0 the concentration at time $t = 0$, then

$$\frac{dC}{dt} = -kC \tag{5.1}$$

where k is a rate constant. This equation has the simple solution

$$C = C_0 e^{-kt} \tag{5.2}$$

and is appropriately known as the exponential decay model because C and dC/dt both decrease exponentially with time. The constant k depends on the drug and can be found experimentally. At time $t = 1/k$, the concentration $C = C_0/e \approx C_0/2.718 \approx 0.37C_0$; so that by this time about 63% of the drug has disappeared from the system. At time $t = 2/k$, $C = C_0/e^2$ so that about 87% of the drug has disappeared and when $t = 3/k$ about 95% of the drug has disappeared from the system.

Suppose now constant doses C_0 are given at equal intervals of time T. We then find that immediately after the first dose the concentration is C_0, immediately after the second dose it is

$$C_1 = C_0 + C_0 e^{-kT}, \tag{5.3}$$

immediately after the third dose it is

$$C_2 = C_0 + \left(C_0 + C_0 e^{-kT}\right) e^{-kT}$$
$$= C_0 + C_0 e^{-kT} + C_0 e^{-2kT}, \tag{5.4}$$

and immediately after the nth dose it is

$$C_{n-1} = C_0 + C_0 e^{-kT} + \cdots + C_0 e^{-(n-1)kT}$$
$$= C_0 \left[1 + e^{-kT} + \cdots + e^{-(n-1)kT}\right]$$
$$= C_0 \frac{1 - e^{-nkT}}{1 - e^{-kT}}. \tag{5.5}$$

As n tends to infinity the concentration approaches an equilibrium value given by

$$C_\infty = \frac{C_0}{1 - e^{-kT}}, \tag{5.6}$$

so that C_∞ represents the ultimate concentration of the drug in the system. The graph of concentration $C(t)$ vs. time is given in Figure 5.1 and shows discontinuities at time $T, 2T, \ldots$ and that the concentration always lies below the line $C(t) = C_\infty$.

At time nT, the concentration due to the first, second, third, $\ldots, n^{\text{th}}$ doses is $C_0 e^{-(n-1)kT}$, $C_0 e^{-(n-2)kT}, \ldots, C_0 e^{-kT}$ and C_0 respectively. Suppose that the first dose is C_1 and that the remaining doses are C_0. Then immediately after the nth dose,

$$C_{n-1} = C_1 e^{-(n-1)kT} + C_0 e^{-(n-2)kT} + \cdots + C_0, \tag{5.7}$$

or, in the limit as $n \to \infty$,

$$C_\infty \to \lim_{n \to \infty} \left\{ C_0 \left[1 + e^{-kT} + \cdots + e^{-(n-2)kT}\right] + C_1 e^{-(n-1)kT} \right\}$$
$$= \frac{C_0}{1 - e^{-kT}}. \tag{5.8}$$

This result is the same as before and leads us to the conclusion that the ultimate concentration will be independent of the initial dose.

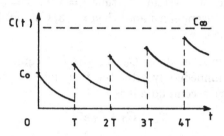

FIGURE 5.1. Exponential decay model for drug concentration

Let us suppose that the initial dose C_1 is equal to the limiting dose C_∞, i.e.,

$$C_1 = C_\infty. \tag{5.9}$$

Then immediately after the first dose the concentration is

$$C_1 = C_\infty = \frac{C_0}{1 - e^{-kT}}, \tag{5.10}$$

immediately after the second dose the concentration is

$$C = C_0 + \left(\frac{C_0}{1 - e^{-kT}} \right) e^{-kT}$$

$$= \frac{C_0}{1 - e^{-kT}}, \tag{5.11}$$

and immediately after the third dose,

$$C = C_0 + \left(\frac{C_0}{1 - e^{-kT}} \right) e^{-kT}$$

$$= \frac{C_0}{1 - e^{-kT}}. \tag{5.12}$$

Thus, the drug concentrations at time 0, T, $2T$, ... are the same. This result is shown in Figure 5.2.

We will now consider the hypothesis that the rate of decrease of the amount of the drug in the system is proportional to the square of the amount, i.e.,

$$\frac{dC}{dt} = -kC^2. \tag{5.13}$$

Integrating and setting $C(t)|_{t=0} = C_0$ (the initial dose) yields

$$C(t) = \frac{C_0}{1 + C_0 kt}. \tag{5.14}$$

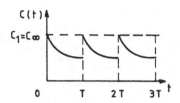

FIGURE 5.2. Drug concentrations in the exponential model when the initial dose is equal to the limiting dose

If after a time $t = T$, a dose C_0 is given, then immediately after the second dose,

$$C_1 = C_0 + \frac{C_0}{1 + C_0 kT}. \tag{5.15}$$

Immediately after the third dose the concentration of the drug in the system is

$$C_2 = C_0 + \frac{C_1}{1 + C_1 kT}. \tag{5.16}$$

Continuing in this way we finally obtain

$$C_n = C_0 + \frac{C_{n-1}}{1 + C_{n-1} kT} \tag{5.17}$$

where C_{n-1} and C_n are the amount of drugs in the system immediately after the nth dose and $(n+1)^{\text{th}}$ dose respectively. Equation (5.17) is a nonlinear difference equation of the first order. An analytical solution of this equation is not possible, but we can deduce some interesting conclusions. From equation (5.17) we can get

$$C_{n+1} = C_0 + \frac{C_n}{1 + C_n kT}. \tag{5.18}$$

Subtracting equation (5.17) from (5.18) gives

$$C_{n+1} - C_n = \frac{C_n - C_{n-1}}{(1 + kTC_n)(1 + kTC_{n-1})}. \tag{5.19}$$

Equaton (5.17) implies that $C_n > C_0$ and from (5.19) we get that $(C_n - C_{n-1})$ and $(C_{n+1} - C_n)$ have the same sign. Also let

$$\lim_{n \to \infty} C_n = C \quad \text{(say)}, \tag{5.20}$$

therefore, from equation (5.17)

$$C = C_0 + \frac{C}{1 + kTC}, \tag{5.21}$$

which leads to

$$C = C_\infty = \frac{kTC_0 + \sqrt{k^2T^2C_0^2 + 4kTC_0}}{2kT}$$
$$= \frac{C_0}{2} + \frac{1}{2}\sqrt{C_0^2 + \frac{4C_0}{kT}}. \tag{5.22}$$

Again this implies that the amount of the drug in the system always lies between C_0 and C_∞.

5.2. Mathematical Modelling in Pharmacokinetics

The mathematical theory of drug phenomena is a branch of the more general mathematical theory of metabolism (a process by which in any living body, human or plants, food is converted to energy). Although drugs are not normal metabolites consumed or produced by the organism, directly or indirectly they do affect different metabolic processes. While a drug is acting, it does take part in some phases of metabolism.

Like the general theory of the metabolism, the theory of drug phenomena presents two important aspects:

(i) The distribution of the drug in the organism;

(ii) The actual biochemical kinetics of the interaction of the drug with different components of the organism and the mechanism of its metabolism.

5.3. The Distribution of Metabolites in the Body

We shall now briefly discuss the problem of distribution of drugs or, in general, of any metabolites in the body.

The organism in the living body has a very heterogeneous structure. It consists of different types of tissues and different fluids surrounding and within these tissues. The various metabolites, whether normal components of the regular metabolism or whether artificially introduced drugs, are transported to various tissues and fluids. These different tissues and fluids may be considered as distinct compartments of the organism. For example, a simple three-compartment structure would be the blood, the intercellular fluid and the cells of some tissue. Exchanges of metabolites take place between these three compartments. The total number of compartments into which an organism may be subdivided is in fact very large. There is also a considerable degree of arbitrariness in the division of the organism into compartments. This division depends very much on the degree of accuracy which is desired in the mathematical model. Thus, if we are satisfied with a rough knowledge of the average amount or concentration of a metabolite in a given organ, we may for our mathematical modelling consider the organ as one compartment. We thus *lump* the cells plus intercellular fluid into one compartment. For a more accurate description we may consider the cells and intercellular fluid as two separate compartments. For a still more accurate theory we will have to take into account that the organ may consist, in general, of several types of cells. In this case, if there are n distinct types of cells in the organ we shall consider $n + 1$ compartments.

Consider now, for modelling purposes, a system of n compartments. Let m different metabolites be involved in the exchange between the compartments. In general these metabolites will also be produced or consumed in some of the compartments. If V_i denotes the volume of the ith compartment, C_{ik} the concentration of the kth metabolite in the ith compartment and M_{ik} the amount of the kth metabolite in the ith compartment, then we have

$$M_{ik} = C_{ik}V_i. \tag{5.23}$$

A direct transport by diffusion, through bounding membranes, from the ith compartment to the jth compartment (say) can occur if the two compartments are adjacent. In this case the influx of the kth metabolite into the ith compartment from the jth one is proportional to the difference in concentration $C_{jk} - C_{ik}$ (according to Rashevsky). If this difference is negative, i.e., $C_{jk} < C_{ik}$, then the metabolite flows from the ith compartment (outflow). If $C_{jk} > C_{ik}$, the flow is from compartment j to i (inflow).

If we denote by $\alpha_{k,ij}$ a factor of proportionality, then the flow or influx is given by

$$\alpha_{k,ij} \left(C_{jk} - C_{ik} \right). \tag{5.24}$$

If the two compartments i and j are not adjacent then $\alpha_{k,ij} = 0$ and there is no flow between them.

The total inflow or outflow of the kth metabolite into or from the ith compartment is given by

$$\sum_{j(\neq i)} \alpha_{k,ij} \left(C_{jk} - C_{ik} \right) \tag{5.25}$$

for fixed i over all values of $j(\neq i)$. If expression (5.25) is positive, the net flow is an inflow, if negative, the net flow is an outflow.

We will now introduce a new parameter, q_{ij}, which denotes the rate of production of the kth metabolite in the ith compartment in $\text{g cm}^{-3}\,\text{s}^{-1}$, such that if $q_{ik} > 0$ then the substance is produced and if $q_{ik} < 0$ then the substance is consumed. The total rate of production is now given by $q_{ik}V_i$ and has units of g s^{-1}. Thus, the net rate of change of the amount M_{ik} is given by

$$\frac{dM_{ik}}{dt} = \sum_{j(\neq i)} \alpha_{k,ij}\left(C_{jk} - C_{ik}\right) + q_{ik}V_i. \qquad (5.26)$$

If both terms on the right side are positive (i.e., $dM_{ik}/dt > 0$) then M_{ik} increases, if these terms are both negative (i.e., $dM_{ik}/dt < 0$) then M_{ik} decreases. When the terms are of opposite sign then the direction of change of M_{ik} depends on the sign of the right side.

Since V_i is constant, from equation (5.23)

$$\frac{dM_{ik}}{dt} = V_i \frac{dC_{ik}}{dt}, \qquad (5.27)$$

which, on comparing with equation (5.26), implies that

$$\frac{dC_{ik}}{dt} = \sum_{j(\neq i)} \beta_{k,ij}\left(C_{jk} - C_{ik}\right) + q_{ik}, \qquad (5.28)$$

where

$$\beta_{k,ij} = \frac{\alpha_{k,ij}}{V_i}. \qquad (5.29)$$

The rate of production q_{ik} is, in general, dependent on the concentration C_{ik}. The simplest possible dependence that is of interest is a simple proportionality, i.e.,

$$q_{ik} = \lambda_{ik}C_{ik}, \qquad (5.30)$$

where λ_{ik} is the coefficient of proportionality. Equation (5.28) can be expanded to give

$$
\begin{aligned}
\frac{dC_{ik}}{dt} &= \sum_{j(\neq i)} \beta_{k,ij}C_{jk} - C_{ik}\sum_{j(\neq i)} \beta_{k,ij} + q_{ik} \\
&= \sum_{j(\neq i)} \beta_{k,ij}C_{jk} - C_{ik}\sum_{j(\neq i)} \beta_{k,ij} + \lambda_{ik}C_{ik} \\
&= \sum_{j(\neq i)} \beta_{k,ij}C_{jk} + \left(\lambda_{ik} - \sum_{j(\neq i)} \beta_{k,ij}\right)C_{ik}. \qquad (5.31)
\end{aligned}
$$

Thus

$$\frac{dC_{ik}}{dt} = \sum_{j=1}^{n} \beta_{k,ij} C_{jk} \qquad (5.32)$$

where

$$\beta_{k,ij} = \alpha_{k,ij}/V_i, \quad \text{for } j \neq i$$

and

$$\beta_{k,ii} = \lambda_{ik} - \sum_{j(\neq i)} \beta_{k,ij}. \qquad (5.33)$$

Since i takes on all values from 1 to n, equation (5.32) represents a system of n linear ordinary differential equations with constant coefficients $\beta_{k,ij}$. The solution of this problem is in general a combination of exponential and sine terms, the exact form of the solution being dependent on the whole set of values of the coefficients $\beta_{k,ij}$. Remember if two compartments i and j are not adjacent then $\beta_{k,ij} = 0$.

5.4. Physiological Application of the Two-Compartment Model

For a two-compartment system with only one metabolite, equation (5.32) reduces to

$$\frac{dC_{i1}}{dt} = \sum_{j=1}^{2} \beta_{1,ij} C_{j1} = \beta_{1,i1} C_{11} + \beta_{1,i2} C_{21} \qquad (5.34)$$

or, for $i = 1$,

$$\frac{dC_{11}}{dt} = \beta_{1,11} C_{11} + \beta_{1,12} C_{21} \qquad (5.35a)$$

and, for $i = 2$,

$$\frac{dC_{21}}{dt} = \beta_{1,21} C_{11} + \beta_{1,22} C_{21}. \qquad (5.35b)$$

In the one-compartment case, equation (5.32) simply gives

$$\frac{dC}{dt} = -kC, \qquad (5.36)$$

where $k = -\beta_{1,11}$ and $C = C_{11}$.

With a slight change of notation we now write equations (5.35) in the form

$$\frac{dx_1}{dt} = -L_{11}x_1 + M_{12}x_2 \qquad (5.37a)$$

and

$$\frac{dx_2}{dt} = M_{21}x_1 - L_{22}x_2 \qquad (5.37b)$$

where $C_{11} = x_1$ and $C_{21} = x_2$. The above equations are known as the fundamental compartment equations for a two-compartment system.

Let us attempt a solution of equations (5.37) in the form

$$x_1(t) = A_1 e^{\lambda t} \qquad (5.38a)$$

and

$$x_2(t) = A_2 e^{\lambda t}. \qquad (5.38b)$$

Upon substituting equation (5.38) into equations (5.37) we obtain

$$(L_{11} + \lambda) A_1 - M_{12} A_2 = 0 \qquad (5.39)$$

and

$$M_{21} A_1 - (L_{22} + \lambda) A_2 = 0 \qquad (5.40)$$

which, for a nontrivial solution, imply that

$$\lambda^2 + (L_{11} + L_{22}) \lambda + (L_{11}L_{22} - M_{12}M_{21}) = 0. \qquad (5.41)$$

As equation (5.41) is a quadratic in λ, there will be two roots, λ_1 and λ_2, such that

$$\lambda_1 + \lambda_2 = - (L_{11} + L_{22}) \qquad (5.42)$$

and

$$\lambda_1 \lambda_2 = L_{11}L_{22} - M_{12}M_{21}. \qquad (5.43)$$

Thus the general solution to the system (5.38) is a linear combination of two linearly independent solutions and can be written as

$$x_1(t) = A_{11}e^{\lambda_1 t} + A_{12}e^{\lambda_2 t} \qquad (5.44a)$$

and

$$x_2(t) = A_{21}e^{\lambda_1 t} + A_{22}e^{\lambda_2 t}. \qquad (5.44b)$$

The initial conditions which prescribe the concentration of x_1 and x_2 at time $t = 0$ imply that

$$x_1(0) = A_{11} + A_{12}, \qquad (5.45a)$$

and

$$x_2(0) = A_{21} + A_{22}. \tag{5.45b}$$

Suppose now that the concentration of a metabolite in compartment 1 (say) has been observed at various times. Denote these observations by $x_1^*(t_j)$, $j = 1, 2, \ldots, N$. Since x_1^* is representable as a sum of two exponentials we must therefore convert the data points $x_1^*(t_j)$ so that they have a representation as a sum of two exponentials. Similarly, we will assume that observations have been made of compartment 2 and that these observations are representable likewise as a sum of two exponentials.

Thus, the constants A_{11}, A_{12}, A_{21}, A_{22}, λ_1 and λ_2 are determinable from observation. From these experimentially determined quantities we wish to infer the properties of the system under investigation. Such a task is inverse to the derivation of the solution we have described above. Instead of deducing the solution from the model, we wish to deduce, from the solution, the values of the parameters of the model, namely L_{11}, L_{22}, M_{12} and M_{21}.

Let us substitute the general solutions (equations (5.44)) into the fundamental compartmental equations (equations (5.37)) and equate the coefficients of $e^{\lambda_1 t}$ and $e^{\lambda_2 t}$. In doing so we obtain the following four equations:

$$-L_{11}A_{11} + M_{12}A_{21} = A_{11}\lambda_1,$$
$$-L_{11}A_{12} + M_{12}A_{22} = A_{12}\lambda_2,$$
$$M_{21}A_{11} - L_{22}A_{21} = A_{21}\lambda_1,$$
$$M_{21}A_{12} - L_{22}A_{22} = A_{22}\lambda_2. \tag{5.46}$$

These, when solved in pairs, yield the following relationships:

$$L_{11} = \frac{-A_{11}A_{22}\lambda_1 + A_{12}A_{21}\lambda_2}{\Delta}, \tag{5.46a}$$

$$M_{12} = \frac{A_{11}A_{12}(\lambda_2 - \lambda_1)}{\Delta}, \tag{5.46b}$$

$$L_{22} = \frac{-A_{11}A_{22}\lambda_2 + A_{12}A_{21}\lambda_1}{\Delta}, \tag{5.46c}$$

$$M_{21} = \frac{A_{21}A_{22}(\lambda_1 - \lambda_2)}{\Delta}, \tag{5.46d}$$

where

$$\Delta = A_{11}A_{22} - A_{12}A_{21}.$$

Thus, the values of the parameters appearing in the compartment equation (5.37) can be determined from the knowledge of the solution.

5.5. Mathematical Modelling of Drug Effects – a More General Approach

We consider the case when a drug is injected intravenously at a constant rate and the drug is taken up by one particular tissue on which it exerts its biochemical reaction.

A two-compartment (blood and tissue) system will now be considered with the parameters for the blood compartment being: the total amount of the drug in the blood x g; the volume of blood v_1; and the concentration of the drug in the blood c_1, while those of the tissue compartment are: the total amount of the drug in the tissue w; the volume of the tissue v_2; and the concentration of the drug in the tissue c_2. Finally the rate of injection is denoted by V g s^{-1}.

We also have that

$$x = c_1 v_1 \ \Rightarrow \ c_1 = \frac{x}{v_1}, \qquad (5.47a)$$

$$w = c_2 v_2 \ \Rightarrow \ c_2 = \frac{w}{v_2}. \qquad (5.47b)$$

The rate of penetration of a drug per cm^2 per second from blood into the tissue will be

$$h\left(c_1 - kc_2\right) \qquad (5.48)$$

where h is the permeability of the membrane which separates the blood from the tissue, and k is a constant known as the partition coefficient for the drug between blood and tissue.

If $c_1 = kc_2$, the two compartments will be in equilibrium and no flow through the membrane will take place. If S denotes the total area of membrane then the total flow from blood to tissue per second will be

$$Sh\left(c_1 - kc_2\right). \qquad (5.49)$$

This will be the rate of loss of the drug from the blood to the tissue. Moreover, normally when a drug is taken it undergoes a decomposition in the blood. We consider for our modelling purpose that the drug decomposes at a rate proportional to its own concentration c_1. Thus we may write the rate of decomposition as

$$k_1 c_1 \qquad (5.50)$$

where k_1 is a constant of proportionality. Hence the rate of disappearance of the drug from the blood due to decomposition is

$$k_1 c_1 v_1, \qquad (5.51)$$

and thus, the total rate of disappearance of the drug from the blood is

$$Sh\,(c_1 - kc_2) + k_1 c_1 v_1. \qquad (5.52)$$

But the total rate of change (dx/dt) of the amount of the drug in the blood is equal to the rate of injection V less the total rate of disappearance. Hence

$$\frac{dx}{dt} = V - \left[Sh\,(c_1 - kc_2) + k_1 c_1 v_1 \right] \qquad (5.53)$$

or, using equations (5.47),

$$\frac{dx}{dt} = V - Sh\left(\frac{x}{v_1} - k\frac{w}{v_2} \right) - k_1 x. \qquad (5.54)$$

We will now introduce two more parameters, namely λ and μ (say), where

$$\lambda = \frac{kv_1}{v_2}, \qquad (5.55a)$$

$$\mu = \frac{Sh}{v_1}. \qquad (5.55b)$$

With the aid of λ and μ equation (5.54) simplifies to

$$\frac{dx}{dt} = V - \mu\,(x - \lambda w) - k_1 x. \qquad (5.56)$$

The amount of drug entering the tissue is equal to the amount leaving the blood, since we considered the case where only one tissue takes up the drug. This amount is given by (5.49), which simplifies to

$$Sh\,(c_1 - kc_2) = \frac{Sh}{v_1}\,(c_1 v_1 - kc_2 v_1) = \mu\,(x - \lambda w). \qquad (5.57)$$

Furthermore, in the tissue the drug is also used up and therefore is gradually destroyed.

Similarly to the above analysis we now assume that the rate of destruction of the drug is proportional to the concentration c_2 in the tissue. Thus we may write the rate of destruction in the tissue using the same constant of proportionality k_1 as

$$k_1 c_2. \qquad (5.58)$$

For the whole tissue, the rate of destruction is $k_1 c_2 v_2$ g s^{-1} which is equal to $k_1 w$ (from equation (5.47b)). Hence the total rate of change (dw/dt) of the amount of the drug in the tissue is

$$\frac{dw}{dt} = \mu\,(x - \lambda w) - k_1 w. \qquad (5.59)$$

The system of differential equations given by equations (5.56) and (5.59) can be readily solved by standard methods. These equations contain three constants, λ, μ and k_1, and express the amounts x and w of the drug in the blood and tissue respectively as a function of time and the three constants. Since their solution has, at present, no practical interest, we will now consider a special case which has some application.

Recalling equations (5.56) and (5.59), namely

$$\frac{dx}{dt} = V - \mu\,(x - \lambda w) - k_1 x, \tag{5.60a}$$

$$\frac{dw}{dt} = \mu\,(x - \lambda w) - k_1 w, \tag{5.60b}$$

on adding we obtain

$$\frac{d}{dt}(x + w) = V - k_1(x + w). \tag{5.61}$$

By multiplying equation (5.60b) by λ and subtracting from equation (5.60a) we obtain

$$\frac{d}{dt}(x - \lambda w) = V - (k_1 + k_2)(x - \lambda w) \tag{5.62}$$

where

$$k_2 = (1 + \lambda)\mu. \tag{5.63}$$

By introducing two new variables as

$$z_1 = x + w, \tag{5.64a}$$

$$z_2 = x - \lambda w, \tag{5.64b}$$

equations (5.61) and (5.62) simplify to

$$\frac{dz_1}{dt} = V - k_1 z_1, \tag{5.65a}$$

$$\frac{dz_2}{dt} = V - (k_1 + k_2)z_2. \tag{5.65b}$$

At the beginning of injection, which we take as time $t = 0$, we have $x = w = 0$ and hence $z_1 = z_2 = 0$ at $t = 0$.

In solving the system (5.65) with the above boundary conditions we get

$$z_1 = \frac{V}{k_1}\left(1 - e^{-k_1 t}\right), \tag{5.66a}$$

and

$$z_2 = \frac{V}{k_1 + k_2}\left(1 - e^{-(k_1 + k_2)t}\right), \tag{5.66b}$$

or, substituting back in $z_1 = x + w$ and $z_2 = x - \lambda w$, we finally obtain

$$w = \frac{V}{1 + \lambda} \left\{ \frac{1}{k_1} \left(1 - e^{-k_1 t}\right) - \frac{1}{k_1 + k_2} \left(1 - e^{-(k_1 + k_2)t}\right) \right\}, \qquad (5.67a)$$

$$x = \frac{V}{1 + \lambda} \left\{ \frac{\lambda}{k_1} \left(1 - e^{-k_1 t}\right) + \frac{1}{k_1 + k_2} \left(1 - e^{-(k_1 + k_2)t}\right) \right\}. \qquad (5.67b)$$

Equation (5.67a) shows that for $t = 0$, $w = 0$, which should realistically be the case, for we began the injection at time $t = 0$ and there was no drug in the system prior to the injection (also this was an initial condition so the result is not surprising). As $t \to \infty$, w asymptotically approaches a constant value given by

$$w_\infty = \frac{V}{1 + \lambda} \left(\frac{1}{k_1} - \frac{1}{k_1 + k_2} \right). \qquad (5.68)$$

The greater the injection rate V, the greater the asymptotic value, w_∞. Since both k_1 and k_2 are positive, we have w_∞ as a positive quantity and as w is proportional to V then as V increases w also increases (see Figure 5.3).

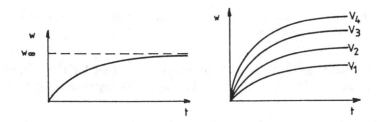

FIGURE 5.3. Relationship of a drug in the tissue with its rate of injection

The action of a drug depends on the instantaneous value of its amount or concentration. The drug exerts some observable action when its amount, or concentration, reaches or exceeds some threshold value w^*.

Let us consider that the action in question is the death of a living body. Then if $w \geq w^*$, the living body dies. If the quantity w_∞ given by equation (5.68) is less than the critical value w^*, then no matter how long we inject the drug, no lethal effect will be observed. In order that this should be the case we must have, from equation (5.68),

$$\frac{V}{1 + \lambda} \left(\frac{1}{k_1} - \frac{1}{k_1 + k_2} \right) < w^* \qquad (5.69)$$

or

$$V < (1 + \lambda)w^* \frac{k_1(k_1 + k_2)}{k_2} = V^* \quad \text{(say).} \tag{5.70}$$

Thus if the rate V of injection is less than V^*, then no matter how long we inject the drug, w will remain less than w^* and no lethal effect will be observed.

On the other hand, if $V > V^*$, then w will reach w^* after a finite time t^* (see Figure 5.4) at t_3^* or t_4^*. This time t^* decreases with increasing V. Obviously when V is very large, t^* will be very small.

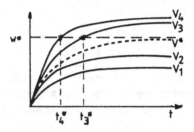

FIGURE 5.4. Relationship of the critical value of a drug with its rate of injection

If we plot t^* against V (see Figure 5.5) we see that for $V = V^*$, we have $t^* \to \infty$, while for $V > V^*$ the value of t^* decreases asymptotically to zero.

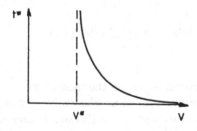

FIGURE 5.5. Relationship of t^* with V

Let us now consider the total dose D_T of the drug injected from $t = 0$ to the point $t = t^*$ at which the lethal threshold is reached. The total dose is equal to

$$D_T = Vt^*. \tag{5.71}$$

The time t^* is obtained by setting the right side of equation (5.67a) equal to w^*. Thus the equation which determines t^* is

$$w^* = \frac{V}{1+\lambda}\left\{\frac{1}{k_1}\left(1 - e^{-k_1 t^*}\right) - \frac{1}{k_1 + k_2}\left(1 - e^{-(k_1+k_2)t^*}\right)\right\}. \qquad (5.72)$$

For a given threshold w^*, this equation gives t^* as a function of V, or V as a function of t^*. This equation cannot be solved explicitly for t^*, though it can be readily solved for V.

Let us investigate the property of $D_T = Vt^*$ without solving equation (5.72). This will lead us to a method of calculating the constants k_1 and k_2 from some observable quantities. Denote in equation (5.72)

$$\left\{\frac{1}{k_1}\left(1 - e^{-k_1 t^*}\right) - \frac{1}{k_1 + k_2}\left(1 - e^{-(k_1+k_2)t^*}\right)\right\} = f(t^*), \qquad (5.73)$$

then equation (5.72) simplifies to

$$w^* = \frac{V}{1+\lambda}f(t^*) \qquad (5.74)$$

and therefore

$$V = \frac{(1+\lambda)w^*}{f(t^*)} \qquad (5.75)$$

or

$$D_T = \frac{(1+\lambda)w^* t^*}{f(t^*)}. \qquad (5.76)$$

We shall now show that the expression $t^*/f(t^*)$ has a minimum value for a certain value t_m^* (say) of t^*. Upon differentiating we get

$$\frac{d}{dt^*}\left(\frac{t^*}{f(t^*)}\right) = \frac{f(t^*) - t^* f'(t^*)}{[f(t^*)]^2}, \qquad (5.77)$$

and for a maximum or minimum value we require

$$f(t^*) - t^* f'(t^*) = 0. \qquad (5.78)$$

Substituting the value of $f(t^*)$ from equation (5.73) and $f'(t^*)$ obtained therefrom we get from equation (5.78)

$$\frac{e^{(k_1+k_2)t_m^*} - 1}{e^{k_1 t_m^*} - 1} = \frac{k_1 + k_2}{k_1}e^{k_2 t_m^*} - (k_1 + k_2)t_m^*\frac{e^{k_2 t_m^*} - 1}{e^{k_1 t_m^*} - 1}. \qquad (5.79)$$

It can be shown that the root of the above equation corresponds to a minimum of the quantity D_T. This equation can readily be solved with the help of a computer.

106 PHARMACOKINETICS: DRUG DISTRIBUTION IN PHARAMACOLOGY

We know that to each value of t^* there corresponds a value of V. Hence to t_m^* there corresponds a value V_m. The equation which determines this value is obtained as follows.

We know that when $V = V^*$, w asymptotically approaches the lethal threshold w^*. Hence we have from equation (5.69) that

$$w^* = \frac{V^*}{1 + \lambda}\left(\frac{1}{k_1} - \frac{1}{k_1 + k_2}\right). \tag{5.80}$$

We then put in equation (5.74) t^* equal to t_m^* and V equal to V_m. Therefore

$$w^* = \frac{V_m}{1 + \lambda}f(t_m^*) \tag{5.81}$$

or from equation (5.80),

$$V^*\left(\frac{1}{k_1} - \frac{1}{k_1 + k_2}\right) = V_m f(t_m^*). \tag{5.82}$$

The value of V^* can be determined experimentally by observing the slowest rate of administration of a drug for which a lethal effect just occurs. Figure 5.6 shows the results from the injection of a drug (digitalis) into cats.

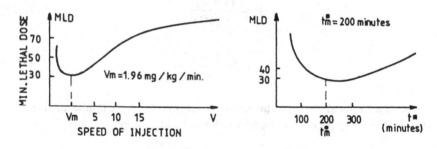

FIGURE 5.6. Results from the injection of a drug into cats

Problems

1. Suppose the concentration of a drug in the blood stream obeys the law

$$\frac{dc}{dt} = -kc^2$$

and equal doses of the drug are given at time $0, T, 2T, 3T, 4T, \ldots$. If c_n is the amount of the drug in the blood immediately after the nth dose, show that the sequence $\{c_n\}$ is monotonically increasing. What is its limit as $n \to \infty$? Find the average amount of the drug in the system in the time interval $(0, nT)$, and find the limit of the average amount as $n \to \infty$.

2. Discuss the general case for the above problem given by

$$\frac{dc}{dt} = -kc^n,$$

where n is a positive integer.

3. A patient is given a dosage Q of a drug at regular intervals of time T. The concentration of the drug in the blood has been shown experimentally to obey the law:

$$\frac{dC}{dt} = -ke^C.$$

(i) If the first dose is administered at $t = 0$ h, show that after T h have elapsed, the residual

$$R_1 = -\ln\left(kT + e^{-Q}\right)$$

remains in the blood.

(ii) Assuming an instantaneous rise in concentration whenever the drug is administered, show that after the second dose and T h have elapsed again, the residual

$$R_2 = -\ln\left\{kT\left(1 + e^{-Q}\right) + e^{-2Q}\right\}$$

remains in the blood.

(iii) Show that the limiting value R of the residual concentrations for doses of Q mg/ml repeated at intervals of T h is given by the formula:

$$R = -\ln\frac{kT}{1 - e^{-Q}}.$$

4. Consider a two-compartment system for which compartment 1 is initially injected with a drug. Subsequent observation of the compartment yields the following expression for the fractional concentration as a function of the time:

$$\frac{x_1(t)}{x_1(0)} = a_1 e^{\lambda_1 t} + a_2 e^{\lambda_2 t},$$

where a_1, a_2, λ_1 and λ_2 are known constants and $a_1 + a_2 = 1$. Show that the upper and lower bounds for the ratio V_2/V_1 are given by

$$\frac{a_1 a_2 (\lambda_1 - \lambda_2)^2}{(-a_2\lambda_1 - a_1\lambda_2)^2} \le \frac{V_2}{V_1} \le \frac{(-a_1\lambda_1 - a_2\lambda_2)^2}{a_1 a_2 (\lambda_1 - \lambda_2)^2},$$

where V_1 and V_2 are the sizes of the two compartments.
   (Hint: Assume in the notation of the text

$$M_{12} = L_{12}V_2/V_1, \quad M_{21} = L_{21}V_1/V_2, \quad L_{21} \le L_{11}$$

and

$$L_{12} \le L_{22}, \quad L_{12}V_2 \le L_{11}V_1, \quad L_{21}V_1 \le L_{22}V_2.)$$

5. Use the above formula to obtain the required bounds for this problem: Ten grams of a drug were administered intravenously at time zero to a human subject. Thereafter, the concentration of the drug in plasma was measured at various times. The concentration was plotted against time on semilog graph paper. The points were fitted by the following equation:

$$x_1(t) = 0.38e^{-1.65t} + 0.18e^{-0.182t}.$$

6. Derive the fundamental compartmental equations for a three-compartment system with a single metabolite. Hence obtain the general solution of the system.

7. Derive the expression given by equation (5.79) in the text.

6

Mathematical Modelling in Epidemiology

6.1. Basic Concepts

In this chapter we will study the spread and control of epidemics through a susceptible population. An epidemic is an unusually large, short-term outbreak of a disease (for example, measles, AIDS, cholera, malaria, etc.). The spread of a disease depends on the mode of transmission, susceptibility, infectious period, resistance and many other factors.

We begin with a population model known as a *simple deterministic model*.

Assume in a given population at time t, $S(t)$ denotes the number of suceptibles, that is, the number of individuals in the population who can be infected, $I(t)$ denotes the number of infected persons in the population, that is, those with the disease who are actively transmitting it, and $R(t)$ denotes the number of individuals removed from the population by recovery, death, immunization or other means. Before looking at specific models, we will make the following assumptions:

(i) The disease is transmitted by contact between the infected individual and a susceptible individual.

(ii) There is no latent period for the disease, hence the disease is transmitted instantaneously on contact.

(iii) All susceptible individuals are *equally* susceptible and all infected individuals are *equally* infectious.

(iv) The population under consideration is fixed in size, that is, no births or migration occur, and all deaths are taken into account.

Hence, if N is the population size, then

$$S(t) + I(t) + R(t) = N = \text{const.} \tag{6.1}$$

6.2. *SI* Model

Consider a simple case in which there are no removals and everyone in the population is either susceptible to the disease or else infected with the disease.

Let S_0 be the initial number of susceptibles in the population in which a number of infected persons, I_0, have been introduced, so that,

$$S(t)\big|_{t=0} = S_0, \qquad I(t)\big|_{t=0} = I_0.$$

Hence, equation (6.1) gives

$$S(t) + I(t) = S_0 + I_0 = \text{const} = N \quad \text{(say)}. \tag{6.2}$$

However, due to infection, the number of susceptibles decreases and the number of infected persons increases.

If we assume that the rate of decrease of $S(t)$, or the rate of increase of $I(t)$, is proportional to the product of the number of susceptibles and the number of infected, then

$$\frac{dS}{dt} = -\alpha SI, \qquad \frac{dI}{dt} = \alpha SI \tag{6.3}$$

where α is a positive constant. Using equation (6.2), equation (6.3) gives us

$$\frac{dS}{dt} = -\alpha S(N - S), \tag{6.4}$$

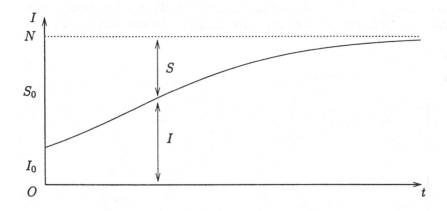

FIGURE 6.1. Limiting behaviour for $S(t)$ and $I(t)$

which is a nonlinear ordinary differential equation. The solution to this equation is given by

$$S(t) = \frac{(N-1)N}{(N-1)+e^{N\alpha t}}. \tag{6.5}$$

Likewise, we obtain

$$I(t) = \frac{Ne^{N\alpha t}}{(N-1)+e^{N\alpha t}}. \tag{6.6}$$

It is informative to look at the limiting behaviour of these solutions. As $t \to \infty$

$$S(t) \to 0, \qquad I(t) \to N. \tag{6.7}$$

Thus, ultimately all persons will be infected (see Figure 6.1).

This suggests that in a large population with a small initial number of infectives, I_0, at first the epidemic (as measured by the total number of infectives) grows exponentially. Then as fewer susceptibles are available, the rate of growth decreases, but the epidemic does not stop until everyone in the population has contracted the disease. Notice that with this model, once an epidemic begins, everyone in the population ultimately contracts the disease. This is because infectives remain infected forever. We will later consider that for most diseases infectives either recover or else they die.

In general, records do not show the number of infected persons but rather the number of new cases reported each day, which is a record of the rate at which the disease is spread in the population. Thus, if we draw a curve of the rate of change in the number of infectives, dI/dt, versus time t, and the rate of change in the number of susceptibles, dS/dt, versus t, remembering $dI/dt = -dS/dt$, then we obtain a curve known as the *epidemic curve*.

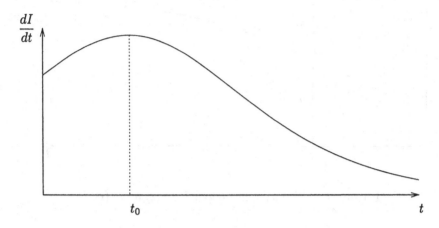

FIGURE 6.2. Epidemic curve

The epidemic curve is a symmetrical unimodal curve with a maximum at

$$t_0 = \frac{\ln(N-1)}{\alpha N}. \tag{6.8}$$

At this point $S = I = N/2$. This shows that the rate of appearance of new cases rises rapidly to its maximum value at a time depending on α and N, and then falls to zero (Figure 6.2).

6.3. SIS Model with Constant Coefficient

In this model, we assume a susceptible person becomes infected at a rate proportional to SI and then an infected person recovers and becomes susceptible at a rate, βI, proportional to the current number of infectives $I(t)$. Hence, our basic equations in this model are given by

$$\frac{dS}{dt} = -\alpha SI + \beta I, \qquad \frac{dI}{dt} = \alpha SI - \beta I. \tag{6.9}$$

We have

$$S(t) + I(t) = \text{const} = S(0) + I(0) = S_0 + I_0 = N \quad \text{(say)}. \tag{6.10}$$

Using equations (6.9) and (6.10), we obtain

$$\frac{dI}{dt} = \alpha(N-I)I - \beta I = (\alpha N - \beta)I - \alpha I^2$$

or

$$\frac{dI}{dt} = \kappa I - \alpha I^2, \quad \text{where } \kappa = \alpha N - \beta. \tag{6.11}$$

The solution to the above equation is found to be

$$I(t) = \begin{cases} \dfrac{e^{\kappa t}}{\dfrac{\alpha}{\kappa}[e^{\kappa t} - 1] + \dfrac{1}{I_0}}, & \kappa \neq 0; \\[2em] \dfrac{1}{\alpha t + \dfrac{1}{I_0}}, & \kappa = 0. \end{cases} \tag{6.12}$$

In the limit as $t \to \infty$, we have

$$I(t) \to \frac{\kappa}{\alpha} = N - \rho, \quad \text{if } \kappa > 0, \text{ i.e., } N > \rho \text{ where } \rho = \frac{\beta}{\alpha} \tag{6.13}$$

and

$$I(t) \to 0, \quad \text{if } \kappa \leq 0, \text{ i.e., } N \leq \rho. \tag{6.14}$$

These results are shown in Figure 6.3.

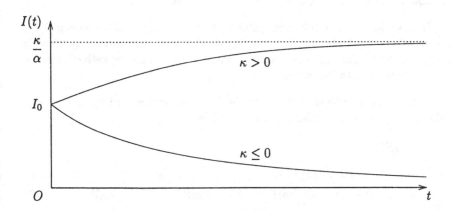

FIGURE 6.3. Infectious curve in SIS Model

6.4. SIS Model when the Coefficient is a Function of Time, t

In this case, we assume $\alpha = \alpha(t)$. Thus, equation (6.11) becomes

$$\frac{dI}{dt} = (\alpha(t)N - \beta)I - \alpha(t)I^2. \tag{6.15}$$

Dividing by I^2, equation (6.15) can be written as

$$\frac{dJ}{dt} + \big(\alpha(t)N - \beta\big)J = \alpha(t) \tag{6.16}$$

where

$$J(t) = \frac{1}{I(t)}. \tag{6.17}$$

Solving the differential equation (6.16), we obtain $I(t)$ in the form

$$I(t) = \frac{\exp\left[N \int_0^t \alpha(u)\,du - \beta t\right]}{\int_0^t \alpha(v)\exp\left[N \int_0^v \alpha(u)\,du - \beta v\right]\,dv + I_0^{-1}}. \tag{6.18}$$

Therefore, depending on the form of $\alpha(t)$, the limiting behaviour of $I(t)$ can be studied in this model.

6.5. *SIS* Model with Constant Number of Carriers

This model is referred to as a *carrier-borne epidemic* model.

Carriers are individuals who, although apparently healthy themselves, harbour infection which can be transmitted to others. Infection may also be derived indirectly through virus contaminated food or liquids, rather than by direct contact with an infective.

In this model, infection is spread both by infectives and by a constant number, C, of carriers so that equation (6.9) becomes

$$\begin{aligned}
\frac{dI}{dt} &= \alpha S(I + C) - \beta I \\
&= \alpha(N - I)(I + C) - \beta I \\
&= -\alpha\left[I^2 - (N - C - \rho)I - CN\right] = -\alpha\left[(I - \alpha_1)(I - \alpha_2)\right].
\end{aligned} \tag{6.19}$$

The solution to (6.19) is given by

$$I(t) = \frac{1}{\alpha}\frac{\alpha_1 a_1 e^{\alpha_1 t} + \alpha_2 a_2 e^{-\alpha_2 t}}{a_1 e^{\alpha_1 t} - a_2 e^{-\alpha_2 t}} \tag{6.20}$$

where

$$\alpha_1, \alpha_2 = \frac{1}{2}\left\{(N - C - \rho) \pm \left[(N - C - \rho)^2 + 4CN\right]^{1/2}\right\} \tag{6.21}$$

such that α_1, α_2 correspond to the positive and negative signs, respectively, and

$$a_1 = \alpha I_0 + \alpha_2, \qquad a_2 = \alpha I_0 - \alpha_1. \tag{6.22}$$

Clearly, in the limit as $t \to \infty$,

$$I(t) \to \frac{\alpha_1}{\alpha} \qquad (6.23)$$

so that $I(t)$ is asymptotic to a positive constant for all values of N and ρ. Thus, with a constant number of carriers, $I(t)$ does not tend to zero, which means the spread of the epidemic is *not controlled.*

6.6. General Deterministic Model with Removal (Kermack–McKendrick Model)

In general, it is difficult to know in advance when a susceptible becomes infected. The existence of the disease only becomes known when symptoms appear. When this occurs, the patient is isolated, and removed from the population. The patient may subsequently either die or recover. As far as transmission of disease is concerned, recovery is a comparatively unimportant event that happens in some cases.

Let the infected persons be removed from the population at a rate proportional to the current number of infective persons. Assuming the former infectives enter a new class which is not susceptible to the disease, we obtain

$$\frac{dS}{dt} = -\alpha SI, \qquad (6.24)$$

$$\frac{dI}{dt} = \alpha SI - \beta I = \alpha I(S - \rho), \qquad \text{where, as before, } \rho = \frac{\beta}{\alpha}, \qquad (6.25)$$

$$\frac{dR}{dt} = \beta I, \qquad \text{where } R \text{ denotes the number of individuals who} \qquad (6.26)$$
$$\text{are isolated, dead or recovered and immune.}$$

Remembering $S + I + R = \text{const} = N$, and with initial conditions at the start of the epidemic

$$S(0) = S_0 > 0, \quad I(0) = I_0 > 0, \quad R(0) = R_0 = 0, \quad S_0 + I_0 = N, \qquad (6.27)$$

we have

$$S(t) + I(t) + R(t) = S_0 + I_0 = N. \qquad (6.28)$$

Unless $\rho < S_0$ (from (6.25)), no epidemic can start to build up, as this requires $(dI/dt)_{t=0} > 0$. Thus $\rho = S_0$ gives a *threshold density of susceptibles.*

Also, from equations (6.24) and (6.26), we have

$$\frac{dS}{dR} = -\frac{S}{\rho} \qquad (6.29)$$

where $\rho = \beta/\alpha$ is called the *relative removal rate*. This is the ratio of the rate at which individuals are removed from the infected category to the rate at which they are added to the same category.

Equation (6.29) yields

$$S = S_0 \exp\left(-\frac{R}{\rho}\right). \tag{6.30}$$

Using equations (6.26), (6.28) and (6.30), we obtain

$$t = \frac{1}{\rho} \int_0^R \frac{dR}{N - S_0 \exp\left(-R/\rho\right) - R}. \tag{6.31}$$

From (6.30) and (6.31) we finally obtain S and R as functions of t and then from equation (6.28) we obtain I as a function of t.

It is clear from equation (6.25) that if $\rho > S_0$, the initial value of $S(t)$, dI/dt, remains negative, indicating that the epidemic does not build up. However, if $\rho < S_0$, that is, the relative removal rate is less than the initial number of susceptibles, the epidemic will build up, and in this case all the susceptible persons do not get infected. Alternatively, if the density of susceptibles is low, the epidemic will disappear.

6.7. Epidemic Model with Vaccination

If in an epidemic an infected person is removed from the population by quarantine, and a susceptible person is made immune by vaccination, then we have the following set of equations:

$$\begin{aligned}
\frac{dS}{dt} &= -\alpha SI - \delta, \\
\frac{dI}{dt} &= \alpha SI - \beta I, \\
\frac{dR}{dt} &= \beta I, \\
\frac{dV}{dt} &= \delta,
\end{aligned} \tag{6.32}$$

where $V(t)$ denotes the number of vaccinated persons, and the initial conditions are

$$R(0) = V(0) = 0.$$

We therefore have

$$S(t) + I(t) + R(t) + V(t) = S_0 + I_0 = N \quad \text{(say)}. \qquad (6.33)$$

The system of equations (6.32) can best be solved by using the technique of dynamic programming. From equation (6.32), we obtain

$$\frac{dI}{dS} = -\frac{\alpha I(S - \rho)}{(\alpha SI + \delta)}. \qquad (6.34)$$

Therefore, if $S > \rho$, dI/dS is negative and if $S < \rho$, dI/dS is positive. The maximum infection occurs when $S = \rho$.

The effect of vaccination on the shape of the trajectories in the SI-phase plane can thus be examined.

Problems

1. Show that the epidemic curve is symmetrical about $t = t_0$.

2. Considering the steady-state solution of the system of equations (6.24)–(6.26), show that

$$R \approx 2\rho(1 - \rho/S_0)$$

   if $I_0 \approx 0$.

3. Substituting $S_0 = \rho + \theta$ in Problem 2 and assuming θ is very small, show

$$R \approx 2\rho \left(\frac{\theta}{\rho + \theta} \right) \approx 2\theta.$$

   This result is known as the *Kermack–McKendrick threshold theorem*, which states that the initial density of susceptibles $\rho + \theta$ is reduced to a final density $\rho - \theta$.

4. In the Kermack–McKendrick model discussed in 6.6, obtain a relation between $S(t)$ and $I(t)$, and interpret the result.

5. Discuss the epidemic model given by

$$\frac{dS}{dt} = -\alpha SI + \mu s, \qquad \frac{dI}{dt} = \alpha SI - \beta I, \qquad \frac{dR}{dt} = \beta I,$$

   which is a general deterministic model. Obtain the trajectories in the SI-plane, and show that the equilibrium position is a stable focus.

7

Modelling the AIDS Epidemic

7.1. A Basic Introduction

AIDS – acquired immunodeficiency syndrome – is a dreadful world epidemic of the 20th century, comparable to the Black Death of the 14th century. The human immunodeficiency virus, HIV, leads to AIDS. When antibodies to HIV are detected, the patient is considered to be infected and is seropositive or HIV positive. The virulence of AIDS and the rate of spread of the epidemic are quite alarming. After the detection of antibodies in the blood, there is a latent period before the end-stage of the disease when the patient exhibits full-blown AIDS. Documented evidence shows that this latent period can be from months to years. In the developed countries, AIDS is primarily associated with the homosexual community, whereas in the underdeveloped regions, the heterosexual spread of the disease is found to be prevalent. The current trends indicate that AIDS is spreading into heterosexual communities of the developed world too.

The reports in the media, journals and international seminars from various countries give a picture of the seriousness and magnitude of the AIDS epidemic. Anderson *et al.* (1986) made a preliminary study of the transmission dynamics of HIV, the causative agent of AIDS. The epidemiology aspects of HIV infection were again studied by Anderson (1988), giving the variable incubation plus infectious periods and heterogeneity in sexual behaviour. We present two simple models from Anderson *et al.* (1986) below, which illustrate how mathematical modelling can be used to understand the dynamics of the evolution of AIDS. For more literature on modelling the AIDS epidemic, the reader may refer to Hahn and Shaw (1986), Murray (1989), Chavez (1989), Blower *et al.* (1991), Hethcote and Van Ark (1992), Weyer and Eggers (1992), Nowak and McLean (1991), Nowak and May (1993), Schulzer *et al.* (1994), and others in recent times.

The major problem with AIDS is the length of the incubation period, that is, from the time the patient is diagnosed as seropositive until symptoms of AIDS appear. In fact, Anderson considered his first model for the time evolution of the disease between those infected with HIV and those with AIDS.

7.2. Anderson's First Model

This simple model was developed by Anderson *et al.* (1986). In this, a population in which all people are infected with HIV is considered, and the time evolution of the disease from the stage of HIV infection to that of AIDS is studied.

Let the size of the HIV infected population who do not have AIDS at time t be $M(t)$, with initial population size $M(0) = M_0$. Let the size of the infected population who have acquired AIDS at time t be $N(t)$, with initial population size $N(0) = 0$. Then the size of the total population under study satisfies

$$M(t) + N(t) = M_0. \qquad (7.1)$$

To nondimensionalize (7.1) divide throughout by M_0. Thus, we obtain

$$x(t) + y(t) = 1 \qquad (7.2)$$

where

$$x(t) = \frac{M(t)}{M_0}, \qquad y(t) = \frac{N(t)}{M_0}. \qquad (7.3)$$

Here $x(t)$ is the fraction of the infected population who do not have AIDS at time t, and $y(t)$ is the fraction of the infected population who have AIDS at time t.

Let $\nu(t)$ be the rate of conversion from infection to AIDS. A simple model for the dynamics with relevant initial condition is

$$\frac{dx}{dt} = -\nu(t)x, \qquad \frac{dy}{dt} = \nu(t)x, \tag{7.4}$$

$$x(0) = 1, \qquad y(0) = 0 \tag{7.5}$$

and

$$x + y = 1. \tag{7.6}$$

In this model, it is assumed that all infected people acquire AIDS (which is not necessarily true). Suppose the victim's immune system gradually breaks down due to the influence of an opportunistic disease like cancer, then the rate at which the infected persons acquire AIDS increases. In other words, $\nu(t)$ is an increasing function of time. Therefore

$$\nu(t) = at, \qquad a > 0, \qquad a \text{ const.} \tag{7.7}$$

Using (7.7), the solution to the system of equations (7.4) is obtained as

$$x(t) = e^{-at^2/2}, \tag{7.8}$$

and

$$y(t) = 1 - e^{-at^2/2}. \tag{7.9}$$

It is clear from (7.8) and (7.9) that

$$\text{as time } t \to \infty, \quad x \to 0 \quad \text{and} \quad y \to 1. \tag{7.10}$$

This result is due to the assumption that all infected people acquire AIDS.

When the data for 194 cases of blood transfusion-associated AIDS presented by Peterman, Drotman and Curran (1981) was compared with the model solution, Anderson et al. (1986) found a best fit value of $a = 0.237$ yr^{-1}. This is shown in Figure 7.1. The parameter a was determined from the best fit to the data. The continuous curve in Figure 7.1 shows the result for the rate of increase dy/dt in AIDS patients as a function of time.

7.3. Anderson's Improved Model

Anderson and May (1989) developed an improved model for AIDS epidemic in a homosexual population, where there is a constant immigration rate of susceptibles. We assume that there is a constant immigration rate B of susceptible

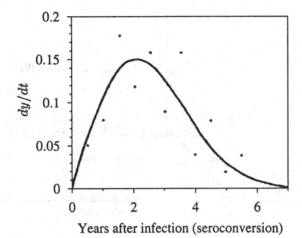

FIGURE 7.1. The rate of change in the proportion of the population who develop AIDS who were infected with HIV (through
blood transfusion) at time $t = 0$

males into a population of size $N(t)$. Let $X(t)$ denote the number of susceptibles at time t, $Y(t)$ the infectious males, $A(t)$ the AIDS patients and $Z(t)$ the
number of seropositives who are not infectious. We assume that susceptibles
die naturally at a rate μ. If there were no AIDS the equilibrium population
would be $N^* = B/\mu$. We assume that AIDS patients die at a rate d (where $1/d$
is of the order of 9 to 12 months). Figure 7.2 is a flow diagram of the disease,
which has been used to formulate the model.

We assume there is uniform mixing in the populations. Based on the flow
diagram of Figure 7.2, a reasonable model can be formulated as

$$\frac{dX}{dt} = B - \mu X - \lambda c X, \tag{7.11}$$

$$\frac{dY}{dt} = \lambda c X - (\nu + \mu)Y, \tag{7.12}$$

$$\frac{dA}{dt} = p\nu Y - (d_\mu)A, \tag{7.13}$$

$$\frac{dZ}{dt} = (1 - p)\nu Y - \mu Z, \tag{7.14}$$

$$N(t) = X(t) + Y(t) + Z(t) + A(t). \tag{7.15}$$

In the above equations B is the immigration rate of susceptibles, μ is the
natural (non-AIDS related) death rate, λ is the probability of acquiring infection from a randomly chosen partner ($\lambda = \beta Y/N$, where β is the transmission

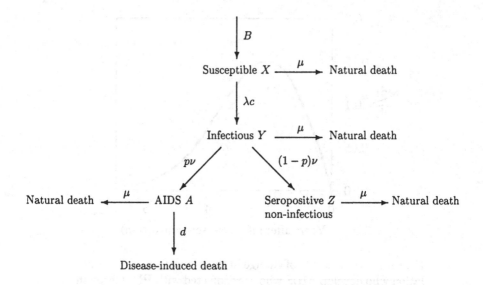

FIGURE 7.2. Flow diagram of the disease for the second model

probability), c is the number of sexual partners, d is the AIDS related death rate, p is the proportion of seropositives who are infectious, and ν (as in the previous model) is the rate of conversion from infection to AIDS, but here taken as a constant. Actually λ here is more appropriately $\beta Y/(X+Y+Z)$, and A is considered small compared to N. With ν constant, $1/\nu$, denoted by D, is then the average incubation time of the disease. Note that in this model the total population $N(t)$ is not a constant, as in the conventional epidemic models. If we add equations (7.11)–(7.14), we obtain

$$\frac{dN}{dt} = B - \mu N - Ad. \tag{7.16}$$

If the basic reproductive rate $R_0 \approx \beta c/\nu > 1$, that is, if the number of secondary infections which arise from a primary infection is greater than one, an AIDS epidemic ensues. From equation (7.12), if at time $t = 0$ an infected individual is introduced into an otherwise infection-free population of susceptibles, remembering initially $X \approx N$ we thus have near $t = 0$,

$$\frac{dY}{dt} \approx (\beta c - \nu - \mu)Y \approx \nu(R_0 - 1)Y, \tag{7.17}$$

since the average incubation time, $1/\nu$, from infection to development of the disease, is far shorter than the average life expectancy, $1/\mu$, of a susceptible, i. e., $\nu \gg \mu$. Thus the approximate threshold condition for the AIDS epidemic to ensue is obtained from equation (7.17) as

$$R_0 \approx \frac{\beta c}{\nu} > 1. \tag{7.18}$$

In the above equation the basic reproductive rate R_0 is given in terms of the number of sexual partners c, the transmission probability β and the average incubation time of the disease $1/\nu$.

After the ensuing of the AIDS epidemic, the system (7.11)–(7.15) evolves to an equilibrium state given by

$$X^* = (\nu + \mu)N^*/C\beta, \tag{7.19}$$

$$Y^* = (d + \mu)(B - \mu N^*)/p\nu d, \tag{7.20}$$

$$Z^* = (1 - p)(d + \mu)(B - \mu N^*)/pd\mu, \tag{7.21}$$

$$A^* = (B - \mu N^*)/d, \tag{7.22}$$

$$N^* = \frac{B\beta\left[(\nu + d + \mu)\mu + \nu d(1 - p)\right]}{(\nu + \mu)[\beta(d + \mu) - p\nu]}. \tag{7.23}$$

We can linearize the system (7.11)–(7.15) about the equilibrium state (7.19)–(7.23), by putting

$$X = X^* + x, \quad Y = Y^* + y, \quad Z = Z^* + z, \quad A = A^* + a, \quad N = N^* + n$$

and it can be proved that x, y, z, a and n tend to zero in an oscillatory manner with a period of oscillation given in terms of the model parameters. We avoid these calculations since they are clumsy. With typical current values for the parameters, the period of epidemic outbreak is of the order of 30 to 40 years. But the parameters characterizing social behaviour associated with the disease may change over this time span.

An interesting item of information from the analysis of the system (7.11)–(7.15) during the early stages of the epidemic is as follows: In the early stages, the population consists almost entirely of susceptibles, and so $X \approx N$, and the equation for the growth of the infectious, that is seropositive, Y class is approximated by the equation (7.17). The solution to (7.17) is

$$Y(t) = Y(0)e^{\nu(R_0-1)t} = Y(0)e^{rt}, \tag{7.24}$$

where R_0 is the basic reproductive rate, $1/\nu$ is the average infectious period and $Y(0)$ is the initial number of infectious people introduced into the susceptible population. The intrinsic growth rate, $r = \nu(R_0 - 1)$, is positive only if an epidemic exists ($R_0 > 1$). From equation (7.24), we can obtain the doubling time for the epidemic, that is the time t_d when $Y(t_d) = 2Y(0)$, as

$$t_d = (\ln 2)/r = (\ln 2)/(\nu(R_0 - 1)). \tag{7.25}$$

From (7.25) we see that the larger the basic reproductive rate R_0 the shorter the doubling time.

If we substitute the solution (7.24) into equation (7.13), for the AIDS patients we get

$$\frac{dA}{dt} = p\nu y(0)e^{rt} - (d + \mu)A. \tag{7.26}$$

Earlier in the epidemic there are no AIDS patients. That is $A(0) = 0$, and so the solution to (7.26) is obtained as

$$A(t) = p\nu Y(0)\frac{e^{rt} - e^{(d+\mu)t}}{r + d + \mu}. \tag{7.27}$$

Estimates for the parameter r have been calculated by Anderson and May (1986) from the data of 6875 homosexual and bisexual men who attended a clinic in San Francisco over the period 1978 to 1985. The average value is obtained as 0.88 per year. Crude estimates (Anderson and May, 1986; Anderson *et al.*, 1986) for values of other parameters are $R_0 = 3$–4, $d + \mu \approx d$, $d = 1$–1.33 per year, $p = 10$–30% or higher, $\nu \approx 0.22$ per year, $c = 2$–6 partners per month. With these estimates we then get an approximate doubling time for the seropositive class as nearly 9 months.

Numerical simulation of the model equations (7.11)–(7.15) gives a clear picture of the epidemic development after the introduction of HIV into a susceptible homosexual population. Figure 7.3 shows one such simulation. The model predicts that the seropositive incidence reaches a maximum around 12 to 15 years after the introduction of the virus into the population.

In this figure the parameter values are $B = 13\,333.3$ yr^{-1}, $\nu = 0.2$ yr^{-1}, $\mu = 1/32$ yr^{-1}, $d = 1$ yr^{-1}, $p = 0.3$, and the basic reproductive rate of the epidemic $R_0 \approx \beta c/\nu = 5.15$. The graphs give the proportion of seropositives and the proportion who develop AIDS. Compare the AIDS curve with that in Figure 7.1.

Even though the above models are simple, they agree with Anderson's (1988) observations made in the homosexual communities. More complex models were proposed by Anderson *et al.* (1986), Anderson (1988), Chavez (1989), Hethcote and Van Ark (1992), Blower *et al.* (1991), Schulzer *et al.* (1994) and others in recent times. With the accumulation of more data and information on the epidemic, more and more sophisticated models are being formed. But the practical use of simple models is that they pose questions which can guide data collection and focus on what useful information can be obtained from sparse data. The crude estimates of epidemic severity doubling time are made with the help of such simple models. Since the epidemic is moving in a major way

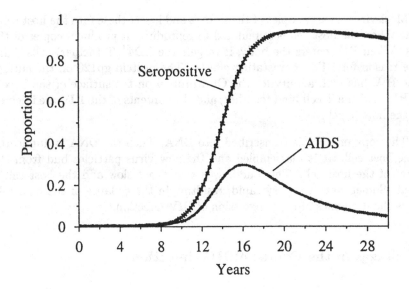

FIGURE 7.3. Numerical simulation of the model system (7.11)–(7.15) with initial conditions $A(0) = Z(0) = 0$, $S(0) + Y(0) = N(0) = 100\,000$

into the heterosexual community in several countries, more and more complex models are required to study the complexity of the epidemic. However, a comprehensive study of various models with complex mathematical formulations is beyond the scope of this chapter.

7.4. Interaction of HIV and Immune System

In this section we use mathematical models and their numerical simulation to understand the interaction of human immunodeficiency virus (HIV) and the human immune system.

As we know, human cells contain a number of chromosomes. The chromosomes become visible in the optical microscope as distinct entities when the cell undergoes division. Chromosomes contain two types of nucleic acid and a number of proteins. The nucleic acids are deoxyribonucleic acid, usually referred to as DNA, and ribonucleic acid or RNA. HIV is a retrovirus in the sense that it carries a copy of its RNA.

Most viruses carry copies of their DNA and insert these into the host cell's DNA. When the host cell is stimulated to reproduce, it produces copies of the virus. When HIV enters the body, it targets the $CD4^+$ T (helper) cells. This is the reason for HIV's devastating effects. The protein gp120 on the surface of the HIV has high affinity for the CD4 protein on the surface of the T cell. The HIV and the T cell bind together, and the contents of the HIV are injected into the host T cell.

The copy of RNA is transcribed into DNA. Then the DNA is duplicated by the host cell. It is reassembled and the new virus particles bud from the surface of the host cell. This budding is sometimes slow and the host cell is spared. Sometimes, it is very rapid, resulting in the collapse of the host cell. This is the dynamics of the progression of HIV infection.

7.5. Stages in the Course of HIV infection

(i) Initial Inoculum

This is initial stage when the virus is introduced into the body.

(ii) Initial Transient

This is a relatively short period of time when both the T cell population and the virus population are in great flux.

(iii) Clinical Latency

This is a period of time when there are extremely large numbers of virus and T cells in the body undergoing incredible dynamics, the overall result of which is an appearance of latency or steady state of the disease.

(iv) AIDS

This stage is characterised by the T cells dropping to very low numbers, even zero, and the virus growing without bound. This ultimately results in death.

7.6. Treatment of HIV Infection

Since HIV is a killer disease, the treatment of HIV infection is of great importance. Several drugs like AZT (zidovudine), DDC, DDI and D4T have been used for treatment, the most widely used drug being AZT. These drugs inhibit the enzyme reverse transcriptase, thereby interfering with the transcription of RNA to DNA. This impedes cellular infection and hence the spread of the virus. Unfortunately, these drugs are not cures. Despite this setback, there is much clinical evidence to support the use of chemotherapy in HIV infected individuals. The advantages are: (i) It can prolong life in the individual, (ii) it may make them less infectious to their partners (Marchuk 1993), and also (iii) reduce the rate of mother-to-foetus transmission (Nozyce *et al.* 1994).

There is a lot of controversial evidence relating to the treatment of HIV infection. Some support the view that early treatment is better, whereas others support treatment at a later stage. There are other questions such as: should the dosage of the drug be large or small, what periodicity of doses should be administered (every four hours, eight hours, etc.).

7.7. Modelling of HIV Immunology

Let us consider a few important cases and their mathematical modelling.

(i) Model 1

A model for the interaction of the immune system with HIV has been formulated by Kirschner (1996).

Variables used in the model

$T(t)$ – uninfected CD4$^+$ T cell population size
$T^i(t)$ – infected CD4$^+$ T cell population size
$V(t)$ – infectious HIV population size
$S(t)$ – source of new CD4$^+$ T cells from the thymus

Parameters and constants used

μ_T – death rate of uninfected CD4$^+$ T cell population
μ_{T^i} – death rate of infected CD4$^+$ T cell population

K_v – rate at which CD4$^+$ T cells become infected by free virus

K_T – rate at which CD4$^+$ T cells kill the virus

r – maximal proliferation of CD4$^+$ T cell population

N – number of free virus produced by bursting infected cells

C – half saturation constant of the proliferation process

b – half saturation constant of the external viral source

g_v – growth rate of external viral source other than T cells

a_{max} – life span of infected CD4$^+$ T cells

a_1 – $[0, a_1]$ is the maximum interval during which reverse transcription occurs

$\gamma(t, a)$ – treatment function, periodic of period p

p – period of dosage in treatment function

c – total daily drug dosage in chemotherapy

k – decay rate of AZT based on half-life of one hour

Construction of Model Equations

(i) Equation for change in the population of uninfected CD4$^+$ T cells

Change in T cells in time dt	=	Increase in T in time dt, due to new T cells from thymus

$$dT \qquad\qquad\qquad\qquad s\,dt$$

	+	Decrease in T in time dt, due to ageing of T cells

(7.28)

$$-\mu_T T dt$$

Increase in T in time dt, due to stimulation of T cells to proliferate in the presence of V	+	Decrease in T in time dt, due to T cells becoming infected by V

$$+\frac{rTV}{C+V}dt \qquad\qquad\qquad -K_v TV dt$$

(ii) Equation for change in the population of infected CD4$^+$ T cells

$$
\boxed{\begin{array}{l}\text{Change in } T^i \text{ in} \\ \text{time } dt, \text{ due to} \\ \text{ageing of } T^i\end{array}} \quad = \quad \boxed{\begin{array}{l}\text{Increase in } T^i \text{ in time} \\ dt, \text{ due to T cells be-} \\ \text{coming infected by } V\end{array}}
$$

$$
dT^i \qquad\qquad\qquad K_v TV dt
$$

$$
+ \boxed{\begin{array}{l}\text{Decrease in } T^i \text{ in time} \\ dt, \text{ due to ageing of } T^i\end{array}} \quad + \quad \boxed{\begin{array}{l}\text{Decrease in } T \text{ in time} \\ dt, \text{ due to stimulation} \\ \text{of T cells to proliferate} \\ \text{in the presence of } V\end{array}} \qquad (7.29)
$$

$$
-\mu_{T^i} T^i dt \qquad\qquad\qquad -\frac{rT^i V}{C + V} dt
$$

(iii) Equation for change in the population of infectious HIV population

$$
\boxed{\begin{array}{l}\text{Change in } V \text{ in} \\ \text{time } dt\end{array}} \quad = \quad \boxed{\begin{array}{l}\text{Increase in } V \text{ in time} \\ dt \text{ as } T^i \text{ cells prolifer-} \\ \text{ate in the presence of} \\ V\end{array}}
$$

$$
dV \qquad\qquad\qquad Nr\frac{T^i V}{C + v} dt
$$

$$
+ \boxed{\begin{array}{l}\text{Decrease in } V \text{ in time} \\ dt, \text{ due to CD8}^+ \text{ T cells} \\ \text{killing } V\end{array}} \quad + \quad \boxed{\begin{array}{l}\text{Increase in } V \text{ in time} \\ dt, \text{ due to growth of} \\ V \text{ from infected cells} \\ \text{other than } T^i\end{array}} \qquad (7.30)
$$

$$
-K_T TV dt \qquad\qquad\qquad +g_v \frac{V}{b + V} dt
$$

Initial conditions

$$
T(0) = T_0,
$$
$$
T^i(0) = T_0^i,
$$
$$
V(0) = V_0. \qquad (7.31)
$$

To include AZT chemotherapy in the model, we include a function $z(t)$, which is "off" outside the treatment period and "on" during the treatment period. This function is given by

$$z(t) = \begin{cases} 1, & \text{outside the treatment period} \\ P(t), & \text{percent effectiveness during AZT treatment} \end{cases} \tag{7.32}$$

where $P(t)$ is a treatment function, $0 < P(t) < 1$.

The function $z(t)$ is multiplied with the parameter K_v in the model. The model now is given by the following system of equations:

$$\frac{dT(t)}{dt} = s(t) - \mu_T T(t) + r\frac{T(t)V(t)}{C + V(t)} - z(t)K_v T(t)V(t) \tag{7.33}$$

$$\frac{dT^i(t)}{dt} = z(t)K_v T(t)V(t) - \mu_{T^i}T^i(t) - r\frac{T^i(t)V(t)}{C + V(t)} \tag{7.34}$$

$$\frac{dV(t)}{dt} = Nr\frac{T^i(t)V(t)}{C + V(t)} + K_T T(t)V(t) + \frac{g_v V(t)}{b + V(t)} \tag{7.35}$$

(ii) Model 2 (Age-structured model)

In the above model, the half-life period of the drug was not taken into account. Moreover, the effect of several doses at different periods of time cannot be studied from this model. In order to overcome these problems, Kirschner (1996) improved the above model by introducing an age structure into it.

Modified variable

$T^i(t, a)$ – density of infected cells with age of cellular infection a, at time t.

The system (7.33)–(7.35) is now modified as follows:

$$\frac{dT(t)}{dt} = s(t) - \mu_T T(t) + r\frac{T(t)V(t)}{C + V(t)} - z(t)K_v T(t)V(t) \tag{7.36}$$

$$T^i(t, 0) = K_v T(t)V(t) \tag{7.37}$$

$$\frac{\partial T^i(t, a)}{\partial t} + \frac{\partial T^i(t, a)}{\partial a} = -\mu_{T^i}T^i(t, a) - r\frac{T^i(t, a)V(t)}{C + V(t)} \tag{7.38}$$

$$\frac{dV(t)}{dt} = Nr\frac{V(t)}{C + V(t)}\int_0^{a_{\max}} T^i(t, a)\, da - K_T T(t)V(t) + \frac{g_v V(t)}{b + V(t)} \tag{7.39}$$

where

$$\int_0^{a_{\max}} T^i(t, a)\, da$$

represents the total infected T cell population at time t, and $a_{\max}$ denotes the maximum age of T cells.

Initial conditions in this case are given by

$$T(0) = T_0,$$
$$V(0) = V_0,$$
$$T^i(0) = T_0^i \qquad \text{for } 0 \le a \le a_{\max}. \tag{7.40}$$

To include AZT chemotherapy in the model, we include a treatment function $\gamma(t, a, p)$. This function is periodic in time, t, with period p and depends on the age a of cellular infection.

The function is given by,

$$\gamma(t, a, p) = \frac{cpe^{-kt}}{1 - e^{-kp}}, \qquad \text{if } 0 \le a \le a_l \text{ and } 0 \le t \le p,$$
$$= \frac{cpe^{-k(t-p)}}{1 - e^{-kp}}, \qquad \text{if } 0 \le a \le a_l \text{ and } p \le t \le 2p,$$
$$\vdots$$
$$= 0, \qquad \text{if } a > a_l \tag{7.41}$$

where k is the decay rate of the drug, p is the periodicity of dosage and c is the intensity of chemotherapy at the beginning of each period.

The model can now be expressed by the following system of equations:

$$\frac{dT(t)}{dt} = s(t) - \mu_T T(t) + r\frac{T(t)V(t)}{C + V(t)} - K_v T(t)V(t) + \int_0^{a_l} \gamma(t, a, p)T^i(t, a)\, dt \tag{7.42}$$

$$T^i(t, 0) = K_v T(t)V(t) \tag{7.43}$$

$$\frac{\partial T^i(t, a)}{\partial t} + \frac{\partial T^i(t, a)}{\partial a} = -\mu_{T^i} T^i(t, a) - r\frac{T(t)V(t)}{C + V(t)} - \gamma(t, a, p)T^i(t, a) \tag{7.44}$$

$$\frac{dV(t)}{dt} = Nr\frac{V(t)}{C + V(t)} \int_0^{a_{\max}} T^i(t, a)\, da - K_T T(t)V(t) + \frac{g_v V(t)}{b + V(t)}. \tag{7.45}$$

Initial conditions in this model are given by

$$T(0) = T_0,$$
$$V(0) = V_0,$$
$$T^i(0) = T_0^i. \tag{7.46}$$

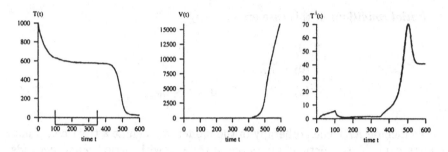

FIGURE 7.4. Graphs of model 1 showing treatment at 100 days (T cells $\sim$ 600 mm^{-3}) for six months

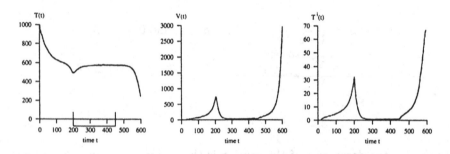

FIGURE 7.5. Graphs of model 1 showing treatment at 200 days (T cells $\sim$ 400 mm^{-3}) for six months

7.8. Analysis of Treatment of HIV Infection

The above models have been used to answer the intriguing questions that arose at the end of Section 7.6.

Results from Model 1

Using the first model (equations (7.33)–(7.35)) we can test different treatment initiations to answer the question whether earlier treatment, initiated after 100 days of infection, or later treatment, initiated after 200 days of infection, is better. Figures 7.4 and 7.5 give a graphical picture of treatment starting at 100 days and 200 days, respectively.

From the results, it seems that the CD4$^+$ T cell count is higher overall when treatment is initiated at the later stages of infection.

Results from Model 2

Using the second model (equations (7.42)–(7.45)) we can simulate treatment to study early versus late treatment. We can also study the periodicity of treatment. Figures 7.6 and 7.7 show three different daily treatment periods: 24 hours, 12 hours and 4.8 hours for an early treatment regime at 100 days and a late treatment regime at 300 days, respectively.

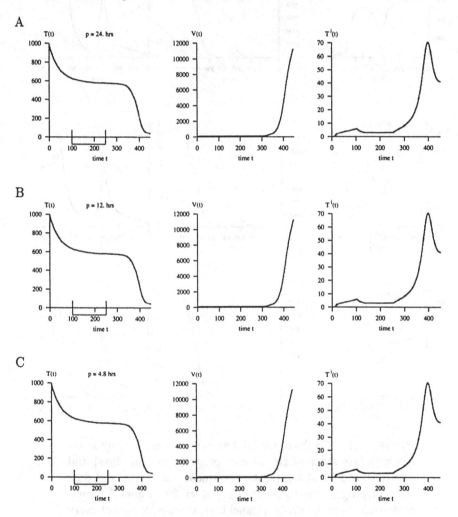

FIGURE 7.6. Graphs of model 2 showing chemotherapy start-ing at an early stage of the disease progression (100 days) and administered for 150 days. All treatment was carried out during the progression to AIDS, i.e., $g_v = 20$. Panel A rep-resents treatment once a day, Panel B represents treatment every twelve hours and Panel C is treatment every 4.8 hours.

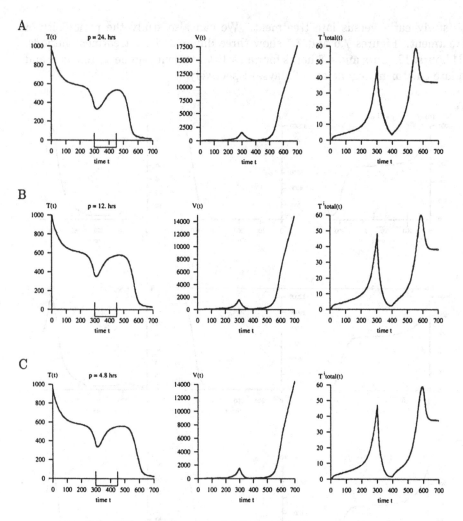

FIGURE 7.7. Graphs of model 2 showing chemotherapy starting at a late stage of the disease progression (300 days) and
administered for 150 days. All treatment was carried out during the progression to AIDS, i.e., $g_v = 20$. Panel A represents treatment once a day, Panel B represents treatment every
twelve hours and Panel C is treatment every 4.8 hours.

From the results two things are evident:

(i) T cell counts are better for later treatment, and

(ii) Period of chemotherapy administration does not affect the overall out-
come of the treatment.

7.9. Conclusion

We have thus arrived at the conclusion that treatment for HIV infection will
be favourable during later stages (200 or 300 days). Mathematical models
given above have answered, to a certain extent, some of the clinical questions
regarding HIV infection.

An enormous amount of money and man-power is being spent on AIDS
research. A number of new drugs are being tested. Besides these, the mathematical community can do a lot of service to humanity by developing improved
and accurate mathematical models for HIV immunology.

8

Biological Fluid Mechanics

8.1. A Basic Introduction

There are many problems in mathematical biology which require a knowledge of fluid flow, for example, the flow of blood through the circulatory system, the swimming of microorganisms, the flight of birds and insects and the motion of bacteria, to name but a few.

Biological science has long been advanced from purely descriptive to analytical science. Many analytical methods of physical science have been used successfully in the study of biological science. Biofluid mechanics does not involve any new development of the general principles of fluid mechanics but it does involve some new applications of the method of fluid mechanics. The most common flow problem in the biological system is the flow of blood. In 1840, the French physician Poiseuille was interested in blood flow and conducted a study of flow in capillaries. It has become known as Poiseuille flow in fluid mechanics, but we know that ordinary Poiseuille flow does not represent the actual blood flow in a cardiovascular system. Many biofluid mechanics problems are concerned with classical fluid mechanics but also its modern aspects such as rheology, chemical reactions, electrothermal effects etc. In this chapter we will outline how we may apply some well-known methods of fluid mechanics to the biological system.

In the physical sciences, the flow of fluid is usually discussed under the heading "fluid dynamics", and the flow of water under the heading "hydrodynamics". In mathematical biology the flow of blood is properly discused under the heading "haemodynamics", which is the study of forces that act on the flow of blood. We now approach the flow of blood from the point of view of the forces which produce this flow.

Approaching the problem from a macroscopic (continuum approach) point of view by assuming that any small volume element of the fluid is so large that it contains a very large number of molecules and also assuming that the fluid is a continuous medium, the dependent variables describing the fluid motion are as follows:

$$\text{pressure } p = p(x, y, z, t),$$
$$\text{density } \rho = \rho(x, y, z, t),$$
$$\text{velocity } \underset{\sim}{q} = \underset{\sim}{q}(x, y, z, t) = (u, v, w).$$

If the density ρ is constant throughout then the flow is said to be *incompressible*. A brief description of viscosity will now be given.

Viscosity

Newton defined the viscosity of a fluid as a lack of slipperiness between the layers of the fluid and, of course, in doing so he implied that there was such a thing as a "layer" or "lamina" of fluid, and the viscosity arises as a result of rubbing one lamina upon the other.

Suppose we have two laminae which are in contact with one another (see Figure 8.1).

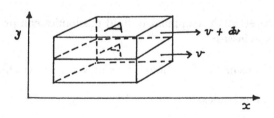

FIGURE 8.1. Newtonian concept of viscosity

Suppose some force F parallel to the x-axis acts and produces relative motion between the two laminae, i.e., the top lamina moves with velocity dv relative to the bottom lamina. Hence, there is a rate of change of velocity with distance in the y direction (i.e., there exists a velocity gradient dv/dy). It is hypothesized that

$$F \propto A \frac{dv}{dy} \tag{8.1}$$

where A is the area of contact between the laminae. The proportionality constant is then defined to be the viscosity of the fluid, and is usually denoted by μ, i.e.,

$$F = \mu A \frac{dv}{dy}. \tag{8.2}$$

The dimensions of the viscosity are given by

$$[\mu] = \frac{[\text{force}]}{[\text{area}] \times \frac{[\text{velocity}]}{[\text{length}]}} = \frac{MLT^{-2}}{L^2 \frac{LT^{-1}}{L}} = \frac{M}{LT}. \tag{8.3}$$

As equation (8.2) is a linear relation, the behaviour of the fluid is called Newtonian. Any fluid for which this relation is nonlinear is called non-Newtonian.

As a result of viscosity, when a viscous fluid flows across a wall, the fluid in immediate contact with the wall is at rest. From above it can be seen that there are many different properties that a fluid may exhibit, and, not surprisingly, Newtonian fluids are often divided up into the following two categories:

(i) Viscous, incompressible;

(ii) Viscous, compressible.

However, the theoretical development of fluid mechanics can primarily be broken down into two classes:

(i) Inviscid fluid mechanics (i.e. nonviscous or viscosity can be neglected);

(ii) Viscous fluid mechanics.

In the next section, we give the basic equations of fluid mechanics to be applied in subsequent sections.

8.2. Basic Equations of Viscous Fluid Motion

If u, v and w denote the velocity in the x, y and z directions respectively, then the basic equations governing fluid motion are:

(a) Continuity (Conservation of Mass)

The continuity equation in vector form is given by

$$\frac{\partial \rho}{\partial t} + \text{div}(\rho \underset{\sim}{q}) = 0. \tag{8.4}$$

If the fluid is incompressible, then $\rho = $ constant and equation (8.4) reduces to

$$\text{div } \underset{\sim}{q} = 0, \tag{8.5}$$

i.e.,

$$\frac{\partial u}{\partial x} + \frac{\partial v}{\partial y} + \frac{\partial w}{\partial z} = 0. \tag{8.6}$$

(b) Navier–Stokes Equations

The vector form of the Navier–Stokes equations is

$$\rho \left(\frac{\partial}{\partial t} + (\underset{\sim}{q} \cdot \nabla) \right) \underset{\sim}{q} = -\nabla p + \mu \nabla^2 \underset{\sim}{q} + \underset{\sim}{F}, \tag{8.7}$$

or in component form in Cartesian coordinates (see Mazumdar, 1992)

$$\rho \left(\frac{\partial u}{\partial t} + u \frac{\partial u}{\partial x} + v \frac{\partial u}{\partial y} + w \frac{\partial u}{\partial z} \right) = -\frac{\partial p}{\partial x} + \mu \nabla^2 u + F_x, \tag{8.8a}$$

$$\rho \left(\frac{\partial v}{\partial t} + u \frac{\partial v}{\partial x} + v \frac{\partial v}{\partial y} + w \frac{\partial v}{\partial z} \right) = -\frac{\partial p}{\partial y} + \mu \nabla^2 v + F_y, \tag{8.8b}$$

$$\rho \left(\frac{\partial w}{\partial t} + u \frac{\partial w}{\partial x} + v \frac{\partial w}{\partial y} + w \frac{\partial w}{\partial z} \right) = -\frac{\partial p}{\partial z} + \mu \nabla^2 w + F_z, \tag{8.8c}$$

where p is the pressure and $\underset{\sim}{F}$ is the external body force vector. As is often the case, if u is the only nonzero velocity component, then the above equations reduce to

$$\frac{\partial \rho}{\partial t} + \frac{\partial (\rho u)}{\partial x} = 0, \tag{8.9a}$$

or if the fluid is incompressible

$$\frac{\partial u}{\partial x} = 0, \tag{8.9b}$$

and

$$\rho\frac{\partial u}{\partial t} = -\frac{\partial p}{\partial x} + \mu\left(\frac{\partial^2 u}{\partial y^2} + \frac{\partial^2 u}{\partial z^2}\right) + F_x. \qquad (8.9c)$$

8.3. Poiseuille's Law (1840)

Poiseuille was a French physician who first investigated in a quantitative manner the flow of water through glass pipes. Poiseuille's interest was the flow of
blood through the vessels of the circulatory system, but he worked with water because of the difficulty at that time of preventing blood from clotting on
exposure to air.

He first considered the flow through a circular cylindrical tube (see Figure
8.2) of length L and radius a, one end of which is at pressure p_1 and the other
end is at pressure p_2, where $p_1 > p_2$ (i.e., a pressure drop). He then derived
an important formula known as Poiseuille's formula for the fluid discharge in
terms of pressure gradient and tube radius.

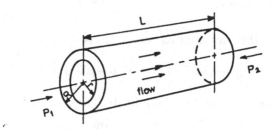

FIGURE 8.2. Flow in a circular cylindrical tube

Assuming the flow is steady (i.e., does not depend on time), and using
cylindrical coordinates (r, θ, x), we seek the solution for the component of the
velocity vector u in the direction of flow. From the symmetry of the problem u
does not depend on the polar angle θ in the plane perpendicular to the x-axis,
so that

$$u = u(x, r). \qquad (8.10)$$

Now from continuity

$$\frac{\partial u}{\partial x} = 0, \qquad (8.11)$$

which implies that

$$u = u(r). \qquad (8.12)$$

From the x component of the Navier–Stokes equations (neglecting body forces), we obtain

$$\rho\left(\frac{\partial u}{\partial t} + u\frac{\partial u}{\partial x}\right) = -\frac{\partial p}{\partial x} + \mu\nabla^2 u, \tag{8.13}$$

which by using equation (8.12) simplifies to

$$\mu\nabla^2 u = \frac{\partial p}{\partial x}. \tag{8.14}$$

The ∇^2 operator in cylindrical coordinates is given by

$$\nabla^2 \equiv \frac{\partial^2}{\partial r^2} + \frac{1}{r}\frac{\partial}{\partial r} + \frac{1}{r^2}\frac{\partial^2}{\partial\theta^2} + \frac{\partial^2}{\partial x^2}. \tag{8.15}$$

Therefore with the help of equation (8.12), equation (8.14) simplifies to

$$\mu\frac{1}{r}\frac{d}{dr}\left(r\frac{du}{dr}\right) = \frac{\partial p}{\partial x}. \tag{8.16}$$

From the r component of the Navier–Stokes equation (8.7), i.e.,

$$\rho\left(\frac{\partial}{\partial t} + (\underset{\sim}{q}\cdot\nabla)\right)q_r = -(\nabla t)_r + \mu\nabla^2 q_r, \tag{8.17}$$

and the fact that u is the only nonzero velocity component, the above implies that

$$\frac{\partial p}{\partial r} = 0, \tag{8.18}$$

which indicates that $p = p(x)$. This leads to the final form for equation (8.16), namely,

$$\mu\frac{1}{r}\frac{d}{dr}\left(r\frac{du}{dr}\right) = \frac{dp}{dx}. \tag{8.19}$$

As u is a function of r only and p is a function of x only, then from a separation of variables argument we conclude that each term of equation (8.19) is constant. Hence,

$$\frac{dp}{dx} = \text{const}, \tag{8.20}$$

which implies that

$$p = p_1 + \frac{p_2 - p_1}{L}x, \tag{8.21}$$

where p_1 is the pressure at $x = 0$ and p_2 is the pressure at $x = L$. Now

$$\frac{d}{dr}\left(r\frac{du}{dr}\right) = \frac{1}{\mu}\left(\frac{dp}{dx}\right)r \tag{8.22}$$

and integration with respect to r gives

$$r\frac{du}{dr} = \frac{1}{\mu}\left(\frac{dp}{dx}\right)\frac{r^2}{2} + A. \tag{8.23}$$

Dividing both sides by r and integrating again yields

$$u = \frac{1}{2\mu}\left(\frac{dp}{dx}\right)\frac{r^2}{2} + A\log r + B. \tag{8.24}$$

Since u at $r = 0$ is finite, we have $A = 0$. Also from the no-slip condition, i.e., $u = 0$ at $r = a$, we have B as

$$B = -\frac{1}{4\mu}\left(\frac{dp}{dx}\right)a^2, \tag{8.25}$$

which gives

$$u = -\frac{1}{4\mu}\frac{dp}{dx}\left(a^2 - r^2\right). \tag{8.26}$$

This expression gives the velocity as zero on the tube's surface and a maximum along the axis. The volume flux (fluid discharge), or the total volume of fluid crossing any section per unit time, is denoted by Q, and is given by

$$Q = \int_0^a 2\pi r u \, dr. \tag{8.27}$$

Upon substitution of equations (8.21) and (8.26), and evaluating the above integral, we obtain

$$Q = \frac{\pi}{8}\frac{(p_1 - p_2)}{\mu L}a^4. \tag{8.28}$$

The above formula for the volume flux is known as Poiseuille's formula or Poiseuille's law, and has the following properties:

(i) It is directly proportional to the fourth power of the radius.

(ii) It is directly proportional to the pressure difference across the tube.

(iii) It is inversely proportional to the length of the tube.

The above results were discovered by Hagen (1839). The "fourth power law" is often invoked to show that if you narrow the tube even slightly the pressure difference required to keep the flow constant must increase greatly. For example, if the arteries become slightly constricted the blood pressure required to supply blood adequately will rise considerably, leading to a state of hypertension.

8.4. Properties of Blood

Poiseuille's law is so well established experimentally that it is often used in order to determine the viscosity coefficient μ of viscous fluids. When blood was examined in this manner it was found that the viscosity of blood $\mu_B = 5\mu_0$ (where μ_0 is the viscosity of water), if the diameter of the tube is relatively large. Thus at normal physiological temperature of 37°C, $\mu_0 = 0.007$ P ($P =$ poise $=$ dynes cm^{-2}) and $\mu_B = 0.035$ P (as determined by Poiseuille's law in large tubes).

The fact that the effective viscosity coefficient of blood according to Poiseuille's law depends on the radius of the tube in which it is measured indicates that blood is not a Newtonian fluid, for which μ is a constant. Rather, blood is said to behave as a non-Newtonian fluid.

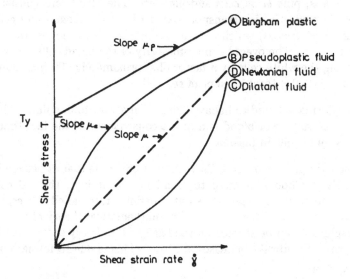

FIGURE 8.3. Typical shear stress–strain rate relationships for non-Newtonian fluids

In the Newtonian case

$$\tau = \mu\dot{\gamma}, \tag{8.29}$$

where τ is the shear stress and $\dot{\gamma}$ is the shear strain rate. A simple model for non-Newtonian behaviour is the power law model given by

$$\tau = \mu\dot{\gamma}^n, \tag{8.30}$$

where n is the power law index. Figure 8.3 shows the relationship between Newtonian and non-Newtonian fluids.

Most of the biological fluids are in fact non-Newtonian. Although adequate mathematical theory of such fluids exists at present, its discussion is beyond the scope of the text. Merchant and Mazumdar (1986), in a study of the non-Newtonian behaviour of blood, observed that the non-Newtonian fluid flow involves many new features not found in the Newtonian fluid flow. However, for most purposes, blood can be treated theoretically as an ordinary Newtonian fluid with an appropriate "effective" viscosity coefficient that is constant.

Blood consists of a suspension of cells in an aqueous solution called *plasma* which consists of 90% water and 7% protein. When blood is subjected to centrifugal force in a centrifuge, a powerful instrument for spinning biochemical solutions at very large angular velocities, it separates out into *plasma* and formed elements, plus *blood cells* and *platelets*. The blood cells consist of red cells (or *erythrocytes*) which transport oxygen from the lungs to all cells, and white cells (or *leukocytes*) which play an important role in the resistance of the body to infections. The overwhelming majority of the blood cells are red cells, outnumbering the white cells by a factor of approximately 600 (i.e., number of white cells = 1/600 × total number of red cells).

Blood plasma is found to behave like a normal Newtonian fluid. Thus, the non-Newtonian nature of blood is a direct consequence of the fact that blood is a *suspension* of cells in plasma.

The specific gravity of red cells is about 1.06, while that of plasma is 1.03. Consequently if blood is allowed to stand in a container, the red cells will settle out of suspension. They do so at a definite rate called the *erythrocyte sedimentation rate* or ESR for short. The mathematical theory of this process is very simple and can be studied on the basis of the theory of fluid mechanics. In illness, the ESR always increases abruptly. This is due to formation of red cell aggregation.

8.5. Application of Poiseuille's Law for the Study of Bifurcation in an Artery

A bifurcation is where the artery branches into two smaller arteries during embryogenesis. The aim is to determine the angle of branching, using the principle of minimization of energy dissipation in conjunction with Poiseuille's law, and also to locate the bifurcation. Firstly, some theory will be given.

It is to be remembered that the problems of blood flow are much more complicated than the problems of fluid flow in engineering situations because of unusually high Reynolds number of flow. The flow remains laminar at Reynolds numbers as high as 5000–10 000 causing the entry length (which is proportional to the Reynolds number) to be very large so that in most cases the fully developed flow is never reached because tube branchings start before this stage is attained. However, we will study this problem with much simplification.

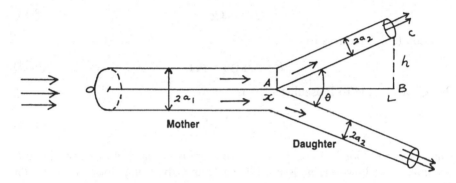

FIGURE 8.4. Determination of the angle of branching using the principle of minimization of energy dissipation and also location of bifurcation.

Law of Conservation of Energy

We consider a mathematical model to predict the pressure changes and volume flow rates in a bifurcating artery. Assume a symmetrical bifurcation when a mother artery is divided into two equal daughter arteries (see Figure 8.4). For steady flow in a tube, if the loss in kinetic energy of the blood in going from the inlet to the outlet is neglected, then the dissipated power P is equal to the work done by the pressure forces at the inlet and outlet ends (according to the law of conservation of energy), which implies that

$$P = Q\Delta p,$$

where P is the dissipated power, Q the flux through the artery, and Δp the pressure drop down its length.

Let p_1 and p_2 be the pressure at the inlet and outlet regions respectively, p_x the pressure at the bifurcated position x, and Q_1 and Q_2 the blood flux in the main artery and bifurcated arteries respectively. The sum of total dissipated power is

$$P = Q_1 (p_1 - p_x) + Q_2 (p_x - p_2) + Q_2 (p_x - p_2) \qquad (8.31)$$

or

$$P = Q_1 (p_1 - p_x) + 2Q_2 (p_x - p_2). \qquad (8.32)$$

Using the conservation of flux, i.e.,

$$Q_1 = 2Q_2, \qquad (8.33)$$

equation (8.32) reduces to

$$P = Q_1 (p_1 - p_2). \qquad (8.34)$$

But Q_1 according to Poiseuille's law is

$$Q_1 = \frac{\pi}{8\mu} \frac{(p_1 - p_x)}{x} a_1^4 \qquad (8.35)$$

where μ is the blood viscosity and a_1 the radius of the main artery. Thus P depends on x; however, p_x is not known. Similarly, using Poiseuille's law for Q_2,

$$Q_2 = \frac{\pi}{8\mu} \frac{(p_x - p_2)}{\sqrt{h^2 + (L - x)^2}} a_2^4. \qquad (8.36)$$

Now by combining (8.33), (8.35) and (8.36) for p_x we obtain

$$p_x = \frac{p_1 a_1^4 \sqrt{h^2 + (L - x)^2} + 2p_2 a_2^4 x}{a_1^4 \sqrt{h^2 + (L - x)^2} + 2a_2^4 x}. \qquad (8.37)$$

Substituting this value of p_x in the expression for P given by equation (8.34) furnishes us P as a function of x, namely,

$$P = \frac{\pi}{4\mu} \frac{a_1^4 a_2^4 (p_1 - p_2)^2}{a_1^4 \sqrt{h^2 + (L - x)^2} + 2a_2^4 x}. \qquad (8.38)$$

The minimum value of P occurs when its first derivative with respect to x is zero keeping all other parameters in (8.38) as constants and $d^2 P / dx^2 > 0$. This happens when

$$x = L - \frac{2a_2^4 h}{\sqrt{a_1^8 - 4a_2^8}}, \qquad (8.39)$$

and hence gives the bifurcation point x.

From Figure 8.4 it can easily be seen that

$$\cos\left(\frac{\theta}{2}\right) = \frac{L-x}{\sqrt{h^2 + (L-x)^2}}, \tag{8.40}$$

or

$$\frac{\theta}{2} = \cos^{-1}\left(\frac{L-x}{\sqrt{h^2 + (L-x)^2}}\right), \tag{8.41}$$

and because x must lie within the range $0 \le x \le L$, then

$$\cos^{-1}\left(\frac{L}{\sqrt{h^2 + L^2}}\right) \le \frac{\theta}{2} \le \frac{\pi}{2}. \tag{8.42}$$

From the combination of equations (8.39) and (8.40) we obtain that for minimum P,

$$\cos\left(\frac{\theta}{2}\right) = 2\left(\frac{a_2}{a_1}\right)^4. \tag{8.43}$$

Thus, from equations (8.42) and (8.43), we arrive at a condition that P will be minimum if

$$2\left(\frac{a_2}{a_1}\right)^4 \ge \frac{L}{\sqrt{h^2 + L^2}}. \tag{8.44}$$

Corresponding to $a_2 = 0$, $x = L$ and $\theta = \pi$, as a_2 increases the bifurcation point x moves to the left and the bifurcation angle θ decreases until $x = 0$, when

$$a_2 = a_1\left(\frac{L^2}{4(h^2 + L^2)}\right)^{1/8} \tag{8.45}$$

and

$$\frac{\theta}{2} = \cos^{-1}\left(\frac{L}{\sqrt{h^2 + L^2}}\right). \tag{8.46}$$

8.6. Pulsatile Flow of Blood

Consider the flow of blood in a circular vessel. Assuming the only non-zero component of the velocity vector is in the axial direction and is denoted by u, using cylindrical polar coordinates, and taking the problem to be axisymmetric (i.e., $u = u(r, x, t)$, $p = p(r, x, t)$), the Navier–Stokes equations reduce to

$$\rho \frac{\partial u}{\partial t} = -\frac{\partial p}{\partial x} + \frac{\mu}{r} \frac{\partial}{\partial r}\left(r \frac{\partial u}{\partial r}\right),$$ (8.47a)

$$0 = -\frac{\partial p}{\partial r}.$$ (8.47b)

The continuity equation gives

$$\frac{\partial u}{\partial x} = 0.$$ (8.47c)

The last two results, equations (8.47b) and (8.47c), imply that

$$p = p(x, t) \quad \text{and} \quad u = u(r, t).$$ (8.48)

Therefore the equations governing the flow simplify to

$$\frac{\mu}{r} \frac{\partial}{\partial r}\left(r \frac{\partial u}{\partial r}\right) - \rho \frac{\partial u}{\partial t} = \frac{\partial p}{\partial x}.$$ (8.49)

Now the left side of the above equation is a function of (r, t) only, the right side a function of (x, t) only, hence both sides must be a function of t only.

Consider now a pulsatile sinusoidal flow with (see Attinger, 1964)

$$\frac{\partial p}{\partial x} = -P e^{i\omega t}$$ (8.50a)

and

$$u(r, t) = U(r) e^{i\omega t}$$ (8.50b)

where P is a constant, and $U(r)$ is the velocity profile across the tube of radius a. We assume that the flow is identical at each section along the tube so that a travelling wave solution can be neglected. It is clear that when $\omega = 0$ (the steady case), the flow becomes that of Poiseuille flow discussed earlier.

From equations (8.50) it is clear that the real part gives the velocity for the pressure gradient $P \cos \omega t$ and the imaginary part gives the velocity for the pressure gradient $P \sin \omega t$. Upon substituting equations (8.50) into equation (8.49) the following equation results:

$$\frac{d^2U}{dr^2} + \frac{1}{r} \frac{dU}{dr} - \frac{i\omega \rho}{\mu} U = -\frac{P}{\mu}.$$ (8.51)

The general solution of an ordinary differential equation in the form

$$\frac{d^2y}{dx^2} + \frac{1}{x} \frac{dy}{dx} - K^2 y = 0$$ (8.52)

is

$$y = A J_0(iKx) + B Y_0(iKx),$$ (8.53)

which involves Bessel functions of complex argument. Thus, the solution of equation (8.51) is

$$U(r) = AJ_0\left(i\sqrt{(i\omega\rho/\mu)}r\right) + BY_0\left(i\sqrt{(i\omega\rho/\mu)}r\right) + \frac{P}{\omega\rho i}. \qquad (8.54)$$

As u and U must be finite on the axis (i.e., at $r = 0$), and since $Y_0(0)$ is not finite, then B has to be zero. Also, because of the no-slip condition $U(r)|_{r=a} = 0$, we have

$$AJ_0\left(i^{3/2}\sqrt{(\omega\rho/\mu)}a\right) + \frac{P}{\omega\rho i} = 0. \qquad (8.55)$$

Let us now introduce a nondimensional parameter α, known as the *Womersley parameter* given by (see also Section 8.10)

$$\alpha = a\sqrt{\frac{\omega\rho}{\mu}} \quad \text{or} \quad \alpha = a\sqrt{\frac{\omega}{\nu}} \qquad (8.56)$$

where $\nu = \mu/\rho$ is the kinematic viscosity. From equation (8.55)

$$A = -\frac{P}{\omega\rho i}\frac{1}{J_0\left(i^{3/2}\alpha\right)} \quad \text{or} \quad A = \frac{iP}{\omega\rho}\frac{1}{J_0\left(i^{3/2}\alpha\right)}, \qquad (8.57)$$

and from equation (8.54)

$$U(r) = -\frac{iP}{\omega\rho}\left(1 - \frac{J_0\left(i^{3/2}\alpha r/a\right)}{J_0\left(i^{3/2}\alpha\right)}\right). \qquad (8.58)$$

In the limit as α approaches zero, i.e., as $\omega \to 0$, the velocity profile becomes parabolic. As α tends to infinity, i.e., viscosity is unimportant ($\mu \to 0$), it can be shown that

$$\frac{J_0\left(i^{3/2}\alpha r/a\right)}{J_0\left(i^{3/2}\alpha\right)} \to 0, \qquad (8.59)$$

which implies that

$$U(r) \to -\frac{iP}{\omega\rho}. \qquad (8.60)$$

In fact, introducing the idea of a "Stokes boundary layer", of thickness δ proportional to $1/\alpha$, it can be concluded that in this case the boundary layer thickness at the cylinder wall disappears.

It is interesting to note that the above expression is independent of viscosity, μ, and exactly 90° out of phase with P. Actually this result corresponds to the Euler equation (which only holds for $\mu = 0$, i.e., inviscid fluids) namely,

$$\frac{\partial u}{\partial t} = -\frac{1}{\rho}\frac{\partial p}{\partial x}. \qquad (8.61)$$

Using equations (8.50) in equation (8.61) we easily see that

$$U = \frac{P}{\rho i \omega} = -\frac{iP}{\rho \omega}, \tag{8.62}$$

which coincides with the result in expression (8.60). Hence the final result for the velocity of pulsatile blood flow in a cylindrical tube of radius a is

$$u(r,t) = -\frac{iP}{\omega \rho} \left(1 - \frac{J_0\left(i^{3/2}\alpha r/a\right)}{J_0\left(i^{3/2}\alpha\right)}\right) e^{i\omega t}. \tag{8.63}$$

The volumetric flow rate Q is given by

$$\begin{aligned}
Q &= \int_0^a u 2\pi r\, dr \\
&= -\frac{2\pi i P}{\omega \rho} e^{i\omega t} \left[\int_0^a r\, dr - \frac{1}{J_0\left(i^{3/2}\alpha\right)} \int_0^a r J_0\left(i^{3/2}\alpha r/a\right) dr\right] \\
&= -\frac{\pi i P}{\omega \rho} e^{i\omega t} a^2 \left[1 - \frac{2}{a^2 J_0\left(i^{3/2}\alpha\right)} \int_0^a r J_0\left(i^{3/2}\alpha r/a\right) dr\right]. \tag{8.64}
\end{aligned}$$

Now consider the term

$$\frac{2}{a^2 J_0\left(i^{3/2}\alpha\right)} \int_0^a r J_0\left(i^{3/2}\alpha r/a\right) dr. \tag{8.65}$$

By making the substitutions

$$i^{3/2}\alpha = \beta \quad \text{and} \quad r\beta/a = \theta$$

the above term simplifies to

$$\frac{2}{\beta^2 J_0\left(\beta\right)} \int_0^\beta \theta J_0\left(\theta\right) d\theta, \tag{8.66}$$

but

$$\int_0^\beta \theta J_0(\theta)\, d\theta = \beta J_1(\beta). \tag{8.67}$$

Therefore (8.66) reduces to

$$\frac{2 J_1(\beta)}{\beta J_0(\beta)} \tag{8.68}$$

and (8.64) can be written as

$$Q = -\frac{P e^{i\omega t}}{\omega \rho} \pi a^2 i \chi(\beta), \tag{8.69}$$

where

$$\chi(\beta) = 1 - \frac{2 J_1(\beta)}{\beta J_0(\beta)}. \tag{8.70}$$

For small values of β (and hence α) the above expression can be expanded in a Taylor series to give

$$\chi(\beta) = 1 - \frac{2}{\beta} \frac{\left(\frac{\beta}{2} - \left(\frac{\beta}{2}\right)^3/1!\,2! + \left(\frac{\beta}{2}\right)^5/2!\,3! - \cdots\right)}{\left(1 - \left(\frac{\beta}{2}\right)^2/(1!)^2 + \left(\frac{\beta}{2}\right)^4/(2!)^2 - \cdots\right)}$$

$$= 1 - \frac{1 - \beta^2/8 + \cdots}{1 - \beta^2/4 + \cdots}$$

$$= 1 - \left(1 - \beta^2/8 + \cdots\right)\left(1 + \beta^2/4 - \cdots\right) = -\beta^2/8 + O(\beta^4). \tag{8.71}$$

By substituting $\beta = i^{3/2}\alpha$ into the above equation we get

$$\chi(\alpha) = \frac{i\alpha^2}{8} + O(\alpha^4). \tag{8.72}$$

Then substituting this result into equation (8.69) and using the definition of α, we finally obtain an expression for Q valid for small α, namely

$$Q = \left(\frac{P\pi a^4}{8\mu} + O(\alpha^4)\right) e^{i\omega t}. \tag{8.73}$$

Again as $\alpha \to 0$ then $\omega \to 0$ and $Q \to Q_0$ where

$$Q_0 = \frac{\pi P a^4}{8\mu} e^{i\omega t} \tag{8.74}$$

and

$$|Q_0| = \frac{\pi P a^4}{8\mu} \tag{8.75}$$

where $|Q_0|$ is the volumetric flow rate for a constant pressure gradient and is the same as for steady Poiseuille's flow.

By separating equation (8.69) into its real and imaginary parts we get

$$Q = \frac{\pi P a^2}{\omega\rho} \left\{ \left[\chi_2(\alpha)\cos\omega t + \chi_1(\alpha)\sin\omega t\right] - i\left[\chi_1(\alpha)\cos\omega t - \chi_2(\alpha)\sin\omega t\right] \right\} \tag{8.76}$$

where

$$\chi(\alpha) = \chi_1(\alpha) + i\chi_2(\alpha).$$

The real part of equation (8.76) gives the flux when the pressure is $P\cos\omega t$ and the imaginary part gives the flux when it is $P\sin\omega t$.

Further Discussion

We are interested in two quantitative aspects of pulsatile flow, namely

(i) the magnitude of Q,

$$|Q| = \frac{\pi Pa^2}{\omega\rho}\sqrt{\chi_1^2(\alpha) + \chi_2^2(\alpha)} = \frac{\pi Pa^4}{\mu\alpha^2}\sqrt{\chi_1^2(\alpha) + \chi_2^2(\alpha)}, \qquad (8.77)$$

(ii) the phase angle between the pressure gradient $Pe^{i\omega t}$ and the flow rate Q, given by (from equation (8.69))

$$\tan\phi = \frac{\chi_1}{\chi_2}. \qquad (8.78)$$

From equation (8.70),

$$\chi_1 + i\chi_2 = 1 - \frac{2}{\alpha i^{3/2}}\frac{J_1\left(i^{3/2}\alpha\right)}{J_0\left(i^{3/2}\alpha\right)}. \qquad (8.79)$$

The graphs of the values of χ_1 and χ_2 from a table of Bessel functions are given in Figure 8.5.

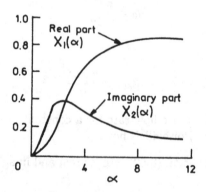

FIGURE 8.5. Graphs of $\chi_1(\alpha)$ and $\chi_2(\alpha)$

We also have

$$\frac{|Q|}{|Q_0|} = \frac{8\sqrt{\chi_1^2 + \chi_2^2}}{\alpha^2}. \qquad (8.80)$$

Since the heart beats about 72 times a minute, we can take

$$\frac{2\pi}{\omega} = \frac{60}{72} \quad \text{or} \quad \omega \approx 8 \text{ rad s}^{-1}.$$

For blood, taking $\rho = 1.05 \text{ g cm}^{-3}$, $\mu = 4 \text{ cP} = 0.04 \text{ g cm}^{-1}\text{s}^{-1}$, and $a = 0.5 \text{ cm}$, we get $\alpha \approx 7$. Even for this value, the volumetric flow rate $|Q|$ is about one-eighth of the steady-state value.

Introducing an impedance parameter z defined by

$$z = \frac{-\frac{\partial p}{\partial x}}{Q} = \frac{Pe^{i\omega t}}{Q}, \tag{8.81}$$

which approaches z_0 as $\alpha \to 0$ (i.e., $\omega \to 0$), where $z_0 = P/Q_0$, we thus have

$$\frac{|z|}{z_0} = \frac{Q_0}{|Q|}. \tag{8.82}$$

Let us now denote a dimensionless quantity defined by

$$\bar{z} = \frac{i\alpha^2}{\chi} = \frac{i\alpha^2}{\chi_1 + i\chi_2} = \frac{\alpha^2 (\chi_2 + i\chi_1)}{\chi_1^2 + \chi_2^2}$$
$$= M + iN \quad \text{(say)}. \tag{8.83}$$

Using equation (8.72) it can be shown that

$$M = 8 + O(\alpha^4) \tag{8.84a}$$

and

$$N = \frac{4}{3}\alpha^2 + O(\alpha^4). \tag{8.84b}$$

For large values of α, by using the asymptotic formula for the Bessel function, it can be shown that

$$M = \sqrt{2\alpha} + 3 + \frac{2\sqrt{2}}{\alpha} + O(\alpha^{-2}) \tag{8.85a}$$

and

$$N = \alpha^2 \left(1 - \frac{\sqrt{2}}{\alpha} + \frac{4\sqrt{2}}{\alpha^2} + O(\alpha^{-3})\right). \tag{8.85b}$$

From equations (8.84), we find that for $\alpha^2 \leq 2/3$ the effect of viscosity forces is more than 90%, and from equations (8.85) we find that for $\alpha^2 \geq 262$ the effect of inertial forces is more than 90%. Substituting $\omega = 8$, $\rho = 1.05 \text{ g cm}^{-3}$, and $\mu = 0.04 \text{ g cm}^{-1}\text{s}^{-1}$ in the expression

$$a^2 = \frac{\mu\alpha^2}{\omega\rho}, \tag{8.86}$$

we obtain $a = 0.65$ mm when $\alpha^2 = 2/3$, and $a = 1$ cm when $\alpha^2 = 162$. Thus, for blood vessels of radius less than 0.65 mm, i.e., for terminal arteries down to microcirculation, inertial effects are small, and most of the energy dissipation in blood flow takes place in this region. For vessels of radius greater than 1 cm, the inertial effects dominate; therefore, in larger arteries, viscous dissipation is of secondary importance. Typical values of artery size, flow rates and Womersley parameters are displayed in Table 8.1.

TABLE 8.1. Relationships of artery size, flow rate and Womersley parameters are shown for the pressure gradient and blood viscosity being taken as $100 \, \mathrm{g \, cm^{-1} \, s^{-2}}$ and $0.04 \, \mathrm{g \, cm^{-1} \, s^{-1}}$ respectively

Womersley Parameter, α	Artery Radius, a (cm)	Flow Rate, $\lvert Q \rvert$ $(\mathrm{cm^3 \, s^{-1}})$
2	0.100	0.083
2	0.200	1.336
2	0.400	21.373
2	0.500	52.180
4	0.100	0.036
4	0.200	0.578
4	0.400	9.251
4	0.500	22.587
6	0.100	0.017
6	0.200	0.276
6	0.400	4.417
6	0.500	10.784
8	0.100	0.010
8	0.200	0.165
8	0.400	2.633
8	0.500	6.429
10	0.100	0.007
10	0.200	0.109
10	0.400	1.737
10	0.500	4.242

The impedance parameter z introduced in equation (8.81) is usually called the longitudinal impedance parameter. The idea of impedance is analogous to the concept of transfer function in a mechanical and electrical system. In fact, if the ratio between the pressure gradient and the flow is used to calculate impedance z as defined by equation (8.81), then it must be characterized by both a magnitude and an angle, and thus can be written as $z = \lvert z \rvert e^{i\theta}$. This phase difference, θ, is very important, since it determines the work done by the system, obtained as the product of pressure and rate of flow. The impedance is a function of position, but not a function of time. If it is used to determine the apparent phase velocity then the speed of propagation will also be a function of position. However, it is highly complex to calculate the impedance as it involves a measurement of both pressure and flow. In practice, the pressure gradient–flow relation has usually been taken from the theoretical equations and used in the reverse direction to calculate pulsatile flow.

8.7. Analysis and Applications of Arterial Flow Dynamics

A comprehensive tutorial on the dynamics of arterial blood flow is beyond the scope of this chapter. For its detailed description, mathematical and engineering science readers may consult any one of the cardiovascular fluid dynamics books, such as by Womersley (1957), McDonald (1974), Rushmer (1976) and Noodergraaf (1978). We will rather emphasize a quantitative way of assessing aspects of arterial blood flow dynamics, which could eventually lead to characterization of normal and diseased states.

The treatment in this section will respond to the following queries:

(i) Can we mathematically demonstrate the generation of pulsatile arterial flow waveform, and what parameters govern this waveform?

(ii) What is the propagation speed of pressure and flow-velocity waveforms?

(iii) How does the arterial constitutive property, i.e., the material property and the geometry of the artery, influence the arterial pressure and pulse wave velocity, and can we express the arterial constitutive property in terms of the measurements of arterial dimensions and pulse wave velocity?

(iv) What is the influence of blood viscosity on the pulse wave propagation analysis, and how can we construct the pressure waveform from the mean flow velocity waveform across the lumen and vice versa?

(v) How can the flow through a stenotic arterial segment be characterized by means of the cyclic time-averaged pressure drop across the stenosed segment, expressed in terms of the time-averaged mean lumen velocity and radius, stenosis length, and degree?

The answers to some of these queries will be given in the following pages.

An especially interesting feature of the circulation is that although the input from the cardiac pump is intermittent, the output from the distributing network is relatively steady. Because of this property, early researchers draw an analogy to the reservoirs of the original fire engine and pipe organs, the *windkessel*, in which compression of air in a chamber provides a stable source of energy for maintaining flow (Noordergraaf, 1978). Accordingly, we begin with this concept, to demonstrate how even this simplistic model can generate a reasonably realistic aortic pressure waveform, in terms of the aortic elasticity, peripheral resistance, aortic inflow rate and diastolic and systolic phase intervals.

Thereafter, we study the analysis for the propagation of pressure pulse waves in an inviscid fluid, to derive the expression for the pulse wave velocity in terms of the diastolic dimensions and elasticity of the artery. The analysis further demonstrates how the pressure and flow velocity waveforms can be expressed in terms of each other, i.e., in the reconstruction of the pressure waveform from the flow velocity waveform and vice versa.

Now the arterial constitutive property governs the deformation response of the artery to a pressure pulse and also the velocity of the pressure pulse. By means of a nonlinear analysis of the arterial vessel wall stress state we have derived expressions for the arterial dimensions and constitutive properties. Therefore, we have demonstrated the method of determination of firstly the arterial constitutive parameters in terms of the diastolic arterial dimensions and pulse wave velocity, and secondly the instantaneous arterial pressure in terms of the instantaneous arterial dimensions, which can yield the possibility of monitoring the cyclic pressure waveform.

We next offer a more representable treatment of arterial pulse wave propagation by allowing for blood viscosity. This yields a characteristic equation in a quantity involving the product of wavespeed and arterial stiffness, and eventually a relationship between the flow velocity wave amplitude and the pressure wave amplitude in terms of the wave frequency, vessel radius and wall thickness, and blood viscosity. By means of this relationship, the arterial pressure waveform is shown to be determinable from the ultrasonically monitorable lumen-averaged velocity amplitude or from the cardiac output, and vice versa.

Finally, a treatment of the flow characteristics in the vicinity of a stenosis is provided. Again, in this analysis, the governing equations are the continuity and Navier–Stokes equations. These equations (upon elimination of the pressure gradient term) yield a single equation, in the centreline axial velocity (with the help of representative velocity profiles), to be solved in terms of the appropriate flow velocity profiles. The pressure drop is determined, and presented in terms of the stenosis degree, stenosis length and Reynolds number (based on time-averaged mean lumen velocity in the unstenosed segment).

8.8. Derivation of Aortic Diastolic–Systolic Pressure Waveforms

In this section, we will develop the analysis for the pressure variation in the aorta during ventricular diastole and systole. This analysis has a physiological significance in enabling us to appreciate the biomedical basis for the shape of the aortic pressure waveform.

It is relevant to provide a brief description of the cardiac cycle and its bearing on the aortic pressure dynamics; for a detailed description of this process, readers may consult any textbook on cardiovascular physiology, such as McDonald (1974) or Rushmer (1976).

During the left ventricular diastolic phase, blood enters the left ventricle through the mitral valve. This filling, along with the onset of ventricular contraction and systole, raises the ventricular pressure above the atrial pressure, whereupon the mitral valve closes. Now, as the ventricle contracts, the pressure builds up in the closed left ventricular chamber until it exceeds the aortic pressure and opens up the aortic valve, thereby initiating the left ventricular ejection phase.

During this phase, the pressure in the aorta builds up. As ejection continues, the left ventricular pressure drops, and the reversal of the ventricular-aortic pressure gradient decelerates the flow until the aortic valve closes. The left ventricle now relaxes, and its pressure drops rapidly until, when it drops below the atrial pressure, the mitral valve opens, the filling phase starts again, and the entire cycle repeats itself.

During the ejection phase, as the blood is pumped into the aorta, the aortic pressure rises and the aorta distends. Thus, not all of the blood pumped into the aorta goes into the systemic circulation, since a portion of it is stored in the distended aorta. The governing equation modulating the aortic pressure is formulated by the consideration that the rate of change of aortic pressure is governed by the product of the volume elasticity (or distensibility) of the aorta and the difference of the outflow rate from the aorta due to ventricular pumping function. The subsequent analysis will illustrate the mathematical implementation of this process.

After closure of the aortic valve, no more blood enters the aorta, but the distended aorta now contracts due to its volume elasticity, propelling blood into the systemic circulation. Thus, the rate of fall of aortic pressure in the elastic aortic chamber during this predominantly cardiac diastolic phase is equal to the product of the volume elasticity of the aorta and the outflow rate, which in turn is equal to the ratio of the aortic pressure and the flow resistance.

We will now set up a mathematical model to describe how these processes govern and shape the pressure waveform in the elastic aortic chamber.

The constitutive relationship between the aortic volume and pressure may be represented by

$$V(t) = V_0 + \frac{P(t)}{K}, \tag{8.87}$$

where K is the volume elasticity of the aorta and V_0 is the end-systolic volume. If the aorta is very elastic, then K is small. We further have

$$\frac{dV}{dt} = I(t) - Q(t), \tag{8.88}$$

where I and Q are inflow and outflow rates, respectively. By combining equations (8.87) and (8.88), we obtain

$$\frac{1}{K}\frac{dP}{dt} = I(t) - Q(t). \tag{8.89}$$

Now we can put down for the outflow rate (by neglecting venous pressure)

$$Q = \frac{P}{R}, \tag{8.90}$$

where R is the effective resistance to aortic flow, depending on the viscosity of the blood and the geometrical parameters of the arteries.

During diastole, the aortic valve is closed and there is no inflow into the aorta. Hence $I(t) = 0$, and we obtain from equations (8.89) and (8.90):

$$\frac{dP}{dt} = -P\frac{K}{R} = -\alpha P, \tag{8.91}$$

where α is a noninvasive measure of the aortic volume elasticity. The solution of the above equation is given by

$$P(t) = P_s e^{-\alpha t}; \qquad 0 \le t \le T_d, \tag{8.92}$$

P_s being the aortic pressure at end-ejection or start of aortic diastole, assuming that the rhythm of the heart beat is quite regular. Equation (8.92) describes the aortic pressure variation during ventricular diastole. If now T_d is the duration of the diastolic phase, then the aortic pressure (P_d), at the end of this phase or just prior to ejection, is given by

$$P_d = P_s e^{-\alpha T_d}. \tag{8.93}$$

This exponential drop of aortic pressure is depicted in Figure 8.6. Equation (8.93) enables us to monitor the volume elasticity of the aorta in terms of end-diastolic and end-systolic pressures P_d and P_s respectively, as

$$\alpha = \frac{1}{T_d} \ln \left(\frac{P_s}{P_d} \right). \tag{8.94}$$

The pressure P_s at the beginning of diastole or, which is the same, at the end of systole is not the same as the systolic pressure P_{systole} which is the maximal pressure during systole. However, P_s is only slightly less than P_{systole}. Therefore, as a first approximation, we may consider P_s as the systolic pressure. Similarly, we may consider P_d as the diastolic pressure.

Equation (8.94) thus affords us, as a byproduct, a nondimensional measure (albeit invasive) of aortic volume elasticity if R remains relatively invariant. We will explain it in detail in the next section.

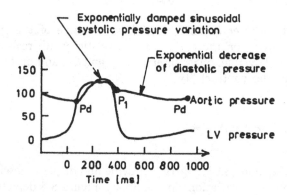

FIGURE 8.6. Tracings of aortic and left ventricular pressures

Now, during systole, the heart pumps blood into the aorta. Let the inflow rate, following Roston (1959, 1962), be represented by

$$I(t) = a \sin \left(\frac{\pi}{T_s} t \right); \qquad 0 \le t \le T_s, \tag{8.95}$$

where a is the maximum value of the inflow rate, and T_s is the duration of the cardiac ejection phase. Thus, from equations (8.89), (8.90) and (8.95), we can put down

$$\frac{1}{K}\frac{dP}{dt} + \frac{P}{R} = a\sin\left(\frac{\pi}{T_s}t\right). \tag{8.96}$$

Now, by imposing the initial condition that at the start of the ejection phase, the aortic pressure $P|_{t=0} = P_d$, we obtain the solution for the pressure variation during the systolic phase as

$$P(t) = P_d e^{-\alpha t} + K a e^{-\alpha t} \int_0^t e^{\alpha t}\sin\beta t\, dt, \tag{8.97}$$

where $\beta = \pi/T_s$ and $P_d = P(t)$ at $t = 0$, i.e., at the start of the ejection or at the end of the aortic diastolic phase.

Upon evaluating the above integral by parts, we obtain the following expression, for the exponentially damped sinusoidal variation of aortic pressure during systole,

$$P(t) = (P_d + A)e^{-\alpha t} + B\sin(\beta t - \psi); \qquad 0 \le t \le T_s \tag{8.98}$$

where

$$A = \frac{a\beta K}{\alpha^2 + \beta^2}, \qquad B = \frac{aK}{\sqrt{\alpha^2 + \beta^2}}, \qquad \psi = \tan^{-1}(\beta/\alpha).$$

It is noted, from equation (8.98), that the value of the systolic aortic pressure is dependent on the aortic volume elasticity, K, resistance to flow, K/α, diastolic and systolic phase intervals, T_d and T_s, and on the inflow rate, a (i.e., on ventricular contractility).

8.9. Clinical Implications

Expressions (8.92) and (8.98) together provide an adequate description of the nature of the pressure variations in the aorta. Figure 8.6 depicts an actually monitored aortic pressure waveform. Note that the diastolic phase pressure variation is an exponential decay, while the systolic phase pressure variation is a damped sinusoidal function, which enables us to appreciate the representativeness of the rather simplistic physics of this derivation.

Measurements of the time period T_d appearing in equation (8.94) of the aortic diastolic pressure can be made by inserting a catheter via the artery of the left arm, the *subclavian*, directly into the aorta and attaching a measuring device to the outside end of the catheter. This type of aortic catheterization experiment was performed by Roston and Leight (1959) who obtained the value of $\alpha\,(= K/R)$ in equation (8.94) as $0.72\,\mathrm{s}^{-1}$. It is known that arteries become less elastic as a person gets older due to the structural changes in the walls of the arteries owing to the cholesterol accumulation, while the lumen of vessels contracts. This contraction of the lumen may cause a myocardial infarction. Thus for clinical purposes, knowledge of the elasticity of the aorta of a patient would be of great significance. Roston and Leight (1959) in fact made this study on the basis of their catheterization experiments performed on 59 subjects between 14 and 58 years of age. They found the volume elasticity of the aorta to lie between 540 and 3870 dyne cm^{-5}. It is interesting to note that the values of volume elasticity of the aorta thus determined for living subjects are of the same order of magnitude as those determined experimentally on cadaver specimens by Remington et al. (1948). However, it must be noted that the aorta is not linearly elastic, as this formulation implies, but is viscoelastic. The resistance R in equation (8.90) and α in equation (8.91) can vary markedly within normal ranges of function, and there are a variety of other nonlinearities (Noodergraaf, 1978; Fung, 1981). Nevertheless, this analysis demonstrates the significant role of arterial compliance in the constitution of central aortic pressure and flow.

8.10. Arterial Pulse Wave Propagation in an Inviscid Fluid

With each heartbeat, blood is pumped into the aorta. Hence the flow of blood in the aorta is pulsatile. At the end of diastole, the fluid in the aorta may be assumed to be at rest. A pressure pulse is then applied to it by the expulsion of blood by the heart into the aorta. This pressure pulse (Figure 8.7a) travels in the fluid-filled aorta and sets up a pressure gradient, as shown in Figure 8.7b.

It is this pressure gradient that determines the resulting pulsatile flow (Figure 8.7c), whose waveform is composed of a steady component (equal to the mean flow during the cardiac cycle period $= (1/T) \int_0^T q(t)\,dt$) and a summation of harmonic oscillating components. The inter-relationship of the pressure and flow wave forms is demonstrated in the ensuing analysis. We will then apply this relationship to demonstrate how the pressure waveform can be mathematically constructed from the monitored lumen-averaged flow waveform.

The propagation speed, c, of the pressure pulse wave is governed, in the first instance, by the elasticity, radius and wall thickness of the aorta, and by the fluid density. The following derived *Moens–Korteweg formula* for the propagation velocity provides the relationship, which will in turn be shown to enable noninvasive assessment of arterial elasticity. The pulse propagation velocity is, in the second instance, also influenced by the blood viscosity. The pulse wave velocity modified by fluid viscosity, c_μ, is pulse frequency dependent, and is actually dependent on a parameter α, the *Womersley parameter* defined

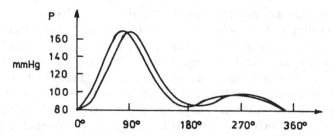

(a) Two pressure waves recorded a short distance
apart in the femoral artery of a dog

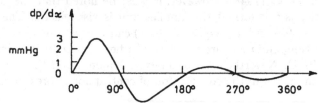

(b) The pressure gradient, derived from the above

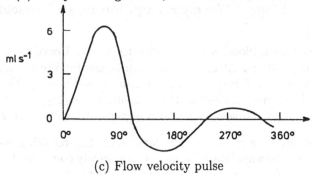

(c) Flow velocity pulse

FIGURE 8.7. Relationships between pulsatile pressure, pressure gradient and flow

earlier. In fact, the parameter α represents the ratio of inertia to viscous forces,
as indicated below

$$\alpha^2 = \frac{\rho\omega a^2}{\mu} = \frac{\rho(\omega u)}{\mu(u/a^2)} = \frac{\rho\dot{u}}{\mu(u/a^2)} \equiv \frac{\text{inertial force per unit volume}}{\text{viscous force per unit volume}},$$

where u is the velocity in the axial direction. At low values of ω and for high
viscosity, μ, $c/c_\mu < 1$. However, for high values of ω and low viscosity (i.e.,
when the inertia force $\gg$ viscous force), c_μ approaches c. The ensuing analysis
will, however, not incorporate the effect of viscosity.

8.11. Moens–Korteweg Expression for Pulse Wave Velocity in an Inviscid Fluid-Filled Elastic Cylindrical Arterial Tube Model

We need to employ the governing fluid-dynamical equations, namely the continuity and momentum equations, the governing arterial tube motion equations and the matching kinematic boundary conditions.

The following assumptions are incorporated in the basic equations:

(i) Blood viscosity is neglected.

(ii) The axial flow velocity is small relative to the pulse wave velocity (of the order of less than 10%).

(iii) The vessel diameter is likewise an order of magnitude smaller than the wavelength.

Let u and v denote the flow velocities in the axial and radial directions (x and r) respectively. By examining each equation separately the following results are obtained:

(a) Continuity Equation

The continuity equation is given by

$$\frac{\partial u}{\partial x} + \frac{1}{r}\frac{\partial(rv)}{\partial r} = 0. \qquad (8.99)$$

We now introduce the average axial flow velocity

$$\bar{u}(x,t) = \frac{1}{\pi a^2}\int_0^a u(x,r,t)2\pi r\,dr, \qquad (8.100)$$

where a is the arterial tube radius at diastole.

Then, upon multiplying both terms of the continuity equation by $2\pi r dr$, integrating with respect to r, and imposing the following boundary conditions:

$$v(x,r,t)|_{r=0} = 0, \qquad \text{(due to axisymmetrical flow)}$$

and

$$v(x,r,t)|_{r=a} = v_\omega, \qquad \text{(wall velocity)}$$

we obtain

$$v_\omega(x,t) = -\frac{a}{2}\frac{\partial \overline{u}(x,t)}{\partial x}. \tag{8.101}$$

(b) Momentum Equations

The momentum equations are

$$\rho\left(\frac{\partial u}{\partial t} + u\frac{\partial u}{\partial x} + v\frac{\partial u}{\partial r}\right) = -\frac{\partial p}{\partial x}, \tag{8.102a}$$

$$\rho\left(\frac{\partial v}{\partial t} + u\frac{\partial v}{\partial x} + v\frac{\partial v}{\partial r}\right) = -\frac{\partial p}{\partial r}. \tag{8.102b}$$

As a result of our assumptions, the radial flow velocity and convective acceleration terms are respectively of smaller order of magnitude than the axial flow velocity and the local acceleration terms. The axial and radial momentum equations now result in

$$\rho\frac{\partial \overline{u}(x,t)}{\partial t} = -\frac{\partial p(x,t)}{\partial x}, \tag{8.103}$$

$$\frac{\partial p}{\partial r} = 0, \tag{8.104}$$

where ρ is the blood density.

(c) Arterial Tube Motion Equation

By considering the dynamic equilibrium of a cylindrical tube element in the radial direction, and by neglecting the inertial force to the motion of the tube wall element (of circumferential and axial dimensions $ad\theta$ and dx respectively, and wall thickness h), we obtain

$$pad\theta dx - (\sigma h dx)d\theta = 0. \tag{8.105}$$

By putting

$$\sigma = E\epsilon = E(\eta/a), \tag{8.106}$$

where η is the wall displacement, and E the wall material elastic modulus, and simplifying, we get from equations (8.105) and (8.106):

$$\eta = \frac{a^2}{hE}p(x,t). \tag{8.107}$$

(d) Kinematic Matching Boundary Conditions

By requiring that the fluid velocity at the wall, v_ω, equals the wall motion velocity ($\dot{\eta}$), and assuming the variation of wall displacement η in the axial direction is negligible, we obtain

$$v_\omega(x,t) = \dot{\eta}(x,t). \tag{8.108}$$

By combining equations (8.101), (8.103), (8.107) and (8.108), we obtain the pulse wave equations, namely,

$$\frac{\partial^2 \bar{u}}{\partial x^2} = \frac{1}{c^2}\frac{\partial^2 \bar{u}}{\partial t^2} \quad \text{and} \quad \frac{\partial^2 p}{\partial x^2} = \frac{1}{c^2}\frac{\partial^2 p}{\partial t^2}, \tag{8.109}$$

where the pulse wave velocity, c, is given by

$$c = \left(\frac{Eh}{2a\rho}\right)^{1/2}. \tag{8.110}$$

8.12. Applications

The above analysis with the derived expression for the pulse wave velocity can provide the basis for characterizing the modulus of elasticity (Young's modulus) as a function of arterial tube's inner radius, as well as for demonstrating the construction of the pressure waveform from the flow velocity waveform and vice versa.

The expression (8.110) for the pulse wave velocity can enable determination of the arterial modulus E (as a function of a) by noninvasive measurement (by sector-scan echocardiography) of the values of arterial dimensions (a, h), the waveforms of the arterial inner radius at two sites, the transit time (as the time interval between the waveform peaks or centroids), and hence the pulse wave velocity. The present-day resolution of sector-scan echocardiography systems can realistically enable us to determine only the aortic pressure with reasonable accuracy.

The solution of equation (8.109) can be written, in d'Alembert's form, as

$$\bar{u} = f(x - ct) + g(x + ct). \tag{8.111}$$

The functions f and g represent waves travelling in the positive and negative x directions respectively with speed c.

Neglecting the reflected wave, and using equation (8.103), we get

$$\bar{u} = f(x - ct); \qquad p = \rho c f(x - ct). \tag{8.112}$$

Thus, the arterial pressure and velocity are proportional to each other; this explains why the pressure and velocity have similar waveforms. This relation can provide the basis of noninvasive measurement of arterial pressure waveform from noninvasive pulsed Doppler flowmeter measurement of the averaged blood flow velocity, $\bar{u}$, waveform, and the noninvasive measurement of the pulse wave velocity as mentioned above.

If the $\bar{u}(x, t)$ waveform is measurable by means of the pulsed Doppler flowmeter, then it can be expresed in Fourier series form by computing the Fourier coefficients, u_n, by means of the Euler integral formula; this is illustrated below.

Assume

$$\bar{u}(x, t) = \sum_{n=-\infty}^{\infty} u_n(x) e^{(i2\pi n t)/T} \qquad (8.113a)$$

such that

$$u_n = \frac{1}{T} \int_0^T \bar{u}(x, t) e^{-(i2\pi n t)/T} \, dt, \qquad (8.113b)$$

where T is the period of the heart beat or cardiac cycle.

Let the corresponding waveform be given by

$$p(x, t) = \sum_{n=-\infty}^{\infty} p_n(x) e^{(i2\pi n t)/T} \qquad (8.114a)$$

where

$$p_n = \frac{1}{T} \int_0^T p(x, t) e^{-(i2\pi n t)/T} \, dt. \qquad (8.114b)$$

Now since from (8.112)

$$p_n = \rho c u_n, \qquad (8.115)$$

we can construct the pressure waveform $p(x, t)$ corresponding to the pulsed Doppler flowmetry derived $\bar{u}(x, t)$ waveform, by following the steps given below:

(i) Determine p_n, associated with u_n of $\bar{u}(x, t)$, corresponding to each constituent frequency parameter $\omega_n = 2\pi n/T$.

(ii) Obtain the composite waveform $p(x, t)$ according to equation (8.114).

(iii) Calibrate the waveform by having the peak waveform amplitude equal to the systolic pressure measured by sphygmomanometry.

Thus, this analysis can enable noninvasive and continuous determination of the pressure waveform. By studying the hitherto unavailable aortic pressure waveforms of different diseases of the cardiovascular system, we can associate waveform parameters with cardiovascular abnormalities and develop a new cardiovascular diagnostic field, based on pressure waveform analysis.

8.13. Towards a Noninvasive Determination of Arterial Pressure from Arterial Wall Stress and Pressure Pulse Propagation Analyses

Let the arterial wall elasticity be represented by

$$E = E_0 e^{k(a-a_0)}, \qquad (8.116)$$

where E_0 and k are the constitutive parameters, and a and a_0 are the deformed and undeformed radii respectively.

From equations (8.110) and (8.116), we have the following expression for the pulse wave velocity c, namely,

$$c^2 = \frac{Eh}{2a\rho} = E_0 e^{k(a-a_0)} \frac{h}{2a\rho}, \qquad (8.117)$$

where h and a are the arterial wall thickness and tube radius respectively, and ρ denotes the blood density.

Now, the arterial wall stress σ, at diastole, is given by

$$\sigma = \frac{pa}{h}. \qquad (8.118)$$

By substituting for σ the constitutive relationship, namely,

$$\sigma = E\epsilon = E\left(\frac{a - a_0}{a_0}\right) \qquad (8.119)$$

we have from equations (8.118) and (8.119) the following expression for the arterial diastolic pressure:

$$p = \sigma\frac{h}{a} = E_0 e^{k(a-a_0)} \left(\frac{a - a_0}{a_0}\right)\left(\frac{h}{a}\right). \qquad (8.120)$$

When the pressure pulse arrives at the arterial site under consideration, the radius a is increased by Δa, the arterial pressure p is increased by Δp, the wall thickness h is decreased by Δh (in order to maintain material incompressibility), the wall stress σ is increased by $\Delta\sigma$, and the strain ϵ is increased by $\Delta\epsilon$, so that

$$\sigma + \Delta\sigma = \frac{(p + \Delta p)(a + \Delta a)}{(h - \Delta h)} = \frac{pa + a\Delta p + p\Delta a + \Delta p\Delta a}{(h - \Delta h)}, \tag{8.121}$$

where the $\Delta p\Delta a$ term is of a smaller order of magnitude than the other terms. By virtue of material incompressibility of the artery, we have

$$(a + \Delta a)(h - \Delta h) = ah. \tag{8.122}$$

From equations (8.121) and (8.122), we then obtain

$$\sigma + \Delta\sigma = \frac{pa}{h} + \frac{a\Delta p}{h} + \frac{2p\Delta a}{h} + \text{higher order terms.} \tag{8.123}$$

Thus, from equations (8.118) and (8.123), we obtain

$$\Delta\sigma = \frac{a\Delta p}{h} + \frac{2p\Delta a}{h}, \tag{8.124}$$

which by invoking equation (8.116) yields

$$\Delta\sigma = E\frac{\Delta a}{a} = E_0 e^{k[(a+\Delta a)-a_0]}\frac{\Delta a}{a} = \frac{a\Delta p}{h} + \frac{2p\Delta a}{h} \tag{8.125}$$

so that

$$E_0 e^{k[(a+\Delta a)-a_0]} = \frac{a^2}{h}\frac{\Delta p}{\Delta a} + \frac{2pa}{h}. \tag{8.126}$$

As we can see, equation (8.120) can enable us to determine the diastolic pressure p (in terms of the ultrasonically monitorable in vivo values of a and h), provided we can first determine the in vivo values of the constitutive parameters (E_0, k) and the undeformed arterial radius a_0. Thereafter, equation (8.126) enables us to determine the systolic pressure increments Δp, and hence the systolic pressure.

To this end, we propose that the pressure pulse velocity be monitored at two different instants $(1, 2)$ following the infusion of either vasoconstrictive or vasodilative drugs that influence the local arterial stiffness. Then, at the two instants, we have

$$c_1^2 = \frac{E_1 h_1}{2a_1\rho} \quad \text{and} \quad c_2^2 = \frac{E_2 h_2}{2a_2\rho}, \tag{8.127}$$

so that by invoking the elasticity expression (8.116), we can write

$$\frac{c_1^2}{c_2^2} = \frac{E_0 e^{k(a_1-a_0)}}{E_0 e^{k(a_2-a_0)}}\frac{h_1}{h_2}\frac{a_2}{a_1} = e^{k(a_1-a_2)}\frac{h_1 a_2}{h_2 a_1}. \tag{8.128}$$

The only unknown variable in the above equation is the constitutive parameter k; the values of all the other terms are noninvasively monitorable and hence known. Equation (8.128) can then be used to determine the value of k. We may then find the values of E_0 and a, using the expressions in (8.127), from the following equations:

$$c_1^2 = \frac{E_0 e^{k(a_1-a_0)} h_1}{2a_1 \rho},$$ (8.129a)

and

$$c_2^2 = \frac{E_0 e^{k(a_2-a_0)} h_2}{2a_2 \rho}.$$ (8.129b)

Having done so, we can now turn to equation (8.120), and determine the value of the diastolic pressure, p. We next proceed to equation (8.126), and determine the value of the systolic incremental pressure, Δp, corresponding to the incremental radius, Δa. Thereby, by ultrasonically monitoring the instantaneous arterial radius, we can incrementally reconstruct the entire systolic pressure waveform.

8.14. Arterial Pulse Wave Propagation Accounting for Blood Viscosity and its Application in Cardiac Output Determination

This theory incorporates the effect of blood viscosity, μ, in the analysis of pulse wave propagation in viscous fluid-filled arterial tubes. The other assumptions stated in Section 8.10 are, however, still relevant.

The blood flow equations of mass continuity and momentum for a viscous fluid are as follows:

$$\frac{\partial u}{\partial x} + \frac{1}{r}\frac{\partial(rv)}{\partial r} = 0,$$ (8.130)

$$\rho\frac{\partial u}{\partial t} + \frac{\partial p}{\partial x} = \mu\left(\frac{\partial^2 u}{\partial r^2} + \frac{1}{r}\frac{\partial u}{\partial r}\right),$$ (8.131)

$$\frac{\partial p}{\partial r} = 0,$$ (8.132)

where u and v are the axial and radial (x and r) blood flow velocities respectively, p is the arterial blood pressure, and ρ is the blood density.

The elastic tube motion equations are (see Fung, 1984):

$$\rho_w h \frac{\partial^2 \eta}{\partial t^2} = \sigma_{rr} - \frac{Eh}{1-\nu^2}\left(\frac{\eta}{a^2} + \frac{\nu}{a}\frac{\partial \xi}{\partial x}\right); \qquad \sigma_{rr} = p, \qquad (8.133)$$

$$\rho_w h \frac{\partial^2 \xi}{\partial t^2} = -\sigma_{rx} - \frac{Eh}{1-\nu^2}\left(\frac{\partial^2 \xi}{\partial x^2} + \frac{\nu}{a}\frac{\partial \eta}{\partial x}\right); \qquad \sigma_{rx} = \mu \frac{\partial u}{\partial r}, \qquad (8.134)$$

where ρ_w denotes the wall density, E the wall modulus, ν is Poisson's ratio, h is the wall thickness, and (ξ, η) are the axial and radial displacements respectively. Further,

$$\left.\frac{\partial \xi}{\partial t}\right|_{r=a} = u; \qquad \left.\frac{\partial \eta}{\partial t}\right|_{r=a} = v. \qquad (8.135)$$

The waveforms for axial and radial blood flow velocities, arterial pressure, and axial and radial wall velocities are represented by:

$$u(x,r,t) = u_0(r)e^{i(kx-\omega t)},$$
$$v(x,r,t) = v_0(r)e^{i(kx-\omega t)},$$
$$p(x,t) = p_0 e^{i(kx-\omega t)}, \qquad (8.136)$$
$$\xi(x,t) = \xi_0 e^{i(kx-\omega t)},$$
$$\eta(x,t) = \eta_0 e^{i(kx-\omega t)},$$

where ω $(= 2\pi/T)$ is the frequency, λ $(= 2\pi/k)$ is the wavelength and ω/k is the wave speed denoted by c.

The actual waveforms will be a composite of harmonic components with different frequencies. However, for the purpose of analysis, it suffices to concern ourselves with a pulse wave of frequency ω, obtain the relationship between the wave amplitudes of the averaged flow velocity and pressure, and thereby demonstrate how a composite pressure wave can be reconstructed from the composite flow velocity wave, which can be monitored by means of pulsed Doppler flowmetry.

Upon substituting the expressions for u and p in equation (8.131), we get the following inhomogeneous Bessel equation in u_0:

$$\frac{d^2 u_0}{dr^2} + \frac{1}{r}\frac{du_0}{dr} + i\omega \frac{\rho u_0}{\mu} = \frac{ikp_0}{\mu}, \qquad (8.137)$$

whose solution is given by

$$u_0(r) = \frac{k}{\omega\rho}p_0 + AJ_0(\beta r) + BY_0(\beta r); \qquad \beta = \frac{i\omega\rho}{\mu}, \qquad (8.138)$$

in which the term $Y_0(\beta r)$ becomes irregular at $r = 0$, and hence must be omitted.

Likewise, we substitute the waveform expressions for u and v in (8.130), integrate the resulting differential equation:

$$\frac{d(rv_0)}{dr} = -ik(ru_0),$$
(8.139)

substitute for $u_0(r)$ from (8.138), and by keeping in mind that $v_0|_{r=0} = 0$ (by virtue of axial symmetry) and that $\int r J_0(r)\, dr = r J_1$, we obtain:

$$v_0(r) = -\frac{ik^2}{2\omega\rho}p_0 r - \frac{ikA}{\beta}J_1(\beta r).$$
(8.140)

Next, we substitute the waveform expressions (8.136) in the dynamic equilibrium equations (8.133) and (8.134) of the arterial wall segment, and obtain

$$p_0 = \frac{Eh}{(1-\nu^2)a^2}\eta_0 + \frac{iEhk\nu}{(1-\nu^2)a}\xi_0 - \rho_w h\omega^2\eta_0,$$
(8.141)

$$\xi_0 = \frac{i\nu}{ak}\eta_0 + \frac{(1-\nu^2)\mu A\beta}{Ehk^2}J_1(\beta a) - \frac{\rho_w h\omega^2(1-\nu^2)}{Ehk^2}\xi_0.$$
(8.142)

The last terms in equations (8.141) and (8.142) are of relatively small order of magnitude, and may thus be neglected.

Next by invoking the boundary conditions (8.135):

$$-i\omega\xi_0 = u_0(a); \qquad -i\omega\eta_0 = v_0(a),$$
(8.143)

where $u_0(a)$ and $v_0(a)$ are given by (8.138) and (8.140), so that

$$\xi_0 = \frac{-k}{i\omega^2\rho}p_0 - \frac{A}{i\omega}J_0(\beta a),$$
(8.144)

$$\eta_0 = \frac{k^2 a p_0}{2\omega^2\rho} + \frac{Ak}{\beta\omega}J_1(\beta a),$$
(8.145)

and by substituting the above equations into equations (8.141) and (8.142), we obtain

$$p_0\left(\frac{E^*}{\rho}\frac{k^2}{\omega^2}\left(\frac{1}{2}-\nu\right)-1\right) + AE^*\frac{k}{\omega}\left(\frac{J_1(\beta,a)}{a\beta} - \nu J_0(\beta a)\right) = 0,$$
(8.146)

and

$$p_0\left(\frac{k}{2\omega(i\omega)\rho}(2-\nu)\right) + \frac{A}{\omega^2}\left(\left(\frac{\mu\beta\omega^2}{E^* ak^2} + \frac{i\omega\nu}{a\beta}\right)J_1(\beta a) + \omega\frac{J_0(\beta a)}{i}\right) = 0,$$
(8.147)

where $E^* = Eh/(1-\nu^2)a$.

For a nonzero wave amplitude, the determinant of the above set of linear equations in p_0 and A must be zero, which finally yields the following characteristic equation in $\overline{X} = (k^2/\omega^2)E^*$, after some mathematical simplification:

$$\left((1 - 2\nu)\overline{X} - 2\rho\right)\left(\left(\frac{\nu}{a\beta} - S\right)\overline{X} + \frac{\rho}{a}\right) + \frac{(2 - \nu)}{2\rho}\left(\frac{1}{a\beta} - S\nu\right)\overline{X}^2 = 0,$$

(8.148a)

where

$$S = \frac{J_0(\beta a)}{J_1(\beta a)}, \quad \text{and} \quad \beta = \frac{i\omega\rho}{\mu}.$$

(8.148b)

From this equation, we can determine the value of the parameter $\overline{X}$ in terms of the known values of a, ν, S and β.

Now ω/k, the wave speed, can be evaluated noninvasively by monitoring the transit time as the time interval between the peaks or centroids of ultrasonically measured waveforms of the arterial diameter at two arterial sites at a known distance apart. Then, since the value of $\overline{X}$ can be determined from equation (8.148a), that of E^* can also be calculated.

Having evaluated E^* and ω/k, we can now compute the value of A in terms of p_0 from either equation (8.146) or (8.147) as

$$A = \frac{-\left(\frac{E^*k^2}{2\rho\omega^2}(1 - 2\nu) - 1\right)p_0}{\frac{E^*k}{\omega}\left(\frac{J_1(\beta a)}{a\beta} - \nu J_0(\beta a)\right)},$$

(8.149)

and thereby express $u_0(r)$ in equation (8.138) completely in terms of p_0 as follows:

$$u_0(r) = \left(\frac{k}{\omega\rho} - \theta J_0(\beta r)\right)p_0,$$

(8.150a)

where

$$\theta = \frac{\left(\frac{E^*k^2(1-2\nu)}{2\rho\omega^2} - 1\right)}{E^*\frac{k}{\omega}\left(\frac{J_1(\beta a)}{a\beta} - \nu J_0(\beta a)\right)}.$$

(8.150b)

The above equation enables the lumen-averaged velocity amplitude, $\overline{u}_0$ (corresponding to the pulse frequency ω), given by

$$\overline{u}_0 = \frac{1}{\pi a^2}\int_0^a u_0(r)2\pi r\, dr,$$

(8.151)

to be related to the corresponding pressure amplitude, p_0, in terms of quantities that are known (such as ρ), monitorable (such as a), and computable (such as S and β). We can now adopt the following steps to construct the arterial pressure waveform:

(i) By means of pulsed Doppler flowmetry, monitor (at an arterial site $x = x_0$), the lumen-averaged flow velocity waveform, namely

$$\bar{u}(x_0, t) = \sum_{n=-\infty}^{\infty} \bar{u}_n e^{in\omega t}. \tag{8.152a}$$

(ii) Determine the spectral components, given by

$$\bar{u}_n = \frac{1}{T} \int_0^T \bar{u}(t) e^{-in\omega t} \, dt. \tag{8.152b}$$

(iii) For each harmonic component $\bar{u}_n$ with frequency $n\omega$, we can determine the associated p_n from equations (8.151) and (8.150a), as

$$p_n = \frac{\pi a^2 \bar{u}_n}{\int_0^a \left(\frac{k}{\omega_n \rho} - \theta J_0(\beta r) \right) 2\pi r \, dr}$$

$$= \frac{a \bar{u}_n}{\left(\frac{ak}{\omega_n \rho} - \frac{2\theta}{\beta} J_1(\beta a) \right)} \tag{8.153}$$

where $\omega_n = n\omega$.

(iv) Obtain the pressure waveform (at a certain $x = x_0$), which is given by

$$p(x_0, t) = \sum_{n=-\infty}^{\infty} p_n e^{i\omega_n t}. \tag{8.154}$$

(v) Finally calibrate $p(x_0, t)$ by making the peak cyclic value equal to the systolic arterial pressure value obtained by means of sphygmomanometry, which amounts to shifting the $p(x_0, t)$ waveform upward by its mean value amount.

We can also express the ascending aortic flow rate, Q, and hence, the cardiac output, CO, in terms of p_0, k and ω, i.e., in terms of the parameters of the ascending aortic pressure waveform.

The cardiac output is given by

$$CO = \int_0^T Q \, dt, \tag{8.155}$$

where

$$Q(t) = \pi a^2 \bar{u}(x_0, t);$$

$$\bar{u}(x_0, t) = \sum_{n=-\infty}^{\infty} \bar{u}_n e^{in\omega t};$$

$$\bar{u}_n = \frac{1}{\pi a^2} \int_0^a u_n(r) 2\pi r \, dr;$$

$$u_n(r) = \left(\frac{k}{\omega_n \rho} - \theta J_0(\beta r) \right) p_n;$$

$$\omega_n = n\omega;$$

$$p_n = \frac{1}{T} \int_0^T p(x_0, t) e^{-i\omega_n t} \, dt.$$

(8.156)

Problems

1. In time-dependent one-dimensional flow of fluid with density ρ and viscosity μ, under pressure gradient G, in a rigid circular tube of radius a, the velocity distribution $u(r, t)$ satisfies

$$\rho \frac{\partial u}{\partial t} = G + \mu \frac{1}{r} \frac{\partial}{\partial r} \left(r \frac{\partial u}{\partial r} \right) \quad \text{for } 0 \le r < a.$$

Here r is the distance from the axis of the tube. What is the boundary condition on $u(r, t)$ at the wall of the tube? Find the solution for velocity distribution in the steady state. Hence prove Poiseuille's law for the volume flow:

$$Q = \frac{\pi G a^4}{8\mu}.$$

Calculate the viscosity of blood for each of the following sets of flow data. Comment on the results.

Radius of Blood Vessel	Length	Mean Velocity of Blood	Pressure Difference
3×10^{-4} cm	0.075 cm	0.25 cm s^{-1}	2×10^4 dyne cm^{-2}
3×10^{-2} cm	3.0 cm	5.0 cm s^{-1}	4×10^3 dyne cm^{-2}

If there is zero pressure gradient, and the initial velocity $u(r, 0)$ has the constant value U over the tube, show that as an approximation valid near the tube wall the equation for $u(r, t)$ can be simplified to the one-dimensional diffusion equation (involving $\nu = \mu/\rho$). Use your knowledge of the error function to find a solution for $u(r, t)$ near the wall, and calculate the shear stress on the wall.

2. Assuming blood to behave like a Newtonian viscous incompressible fluid, show that for the unsteady flow of blood in a circular tube the governing equations and associated boundary conditions are given by

$$\rho \frac{\partial u}{\partial t} = G + \mu \frac{1}{r} \frac{\partial}{\partial r} \left(r \frac{\partial u}{\partial r} \right),$$

$$u = 0 \quad \text{at } t = 0,$$

$$u = 0 \quad \text{at } r = R,$$

and

$$u \text{ is finite at } r = 0.$$

Using separation of variables or otherwise, show that the solution to this boundary value problem in the unsteady state is

$$u(r, t) = \frac{GR^2}{4\mu} \left(1 - \frac{r^2}{R^2} \right) - 8 \sum_{n=1}^{\infty} \frac{J_0 \left(k_n r / R \right)}{k_n^3 J_1(k_n)} e^{-\nu k_n^2 t / R^2},$$

where $\nu = \mu / \rho$.

3. Consider the pulsatile flow of blood in an artery of circular cross section. Assume the pressure gradient $(\partial p / \partial x)$ and axial velocity component of flow $u(r, t)$ satisfy

$$\rho \frac{\partial u}{\partial t} = -\frac{\partial p}{\partial x} + \frac{\mu}{r} \frac{\partial}{\partial r} \left(r \frac{\partial u}{\partial r} \right).$$

Obtain the velocity profile when the flow is pulsatile sinusoidal flow. Discuss the cases when the Womersley parameter has the values 0 and ∞.

4. A main vessel, cross-sectional area A_1, divides into n branches of equal size, each of cross-sectional area A_2.

   (i) Use the principle of conservation to calculate the ratio of mean velocities before and after the branching. (Write $d = nA_2/A_1$, the factor by which the total cross-sectional area changes).

   (ii) Assuming Poiseuille's law, find the ratio of pressure gradient in the main vessel to that of the branches.

   (iii) What value of d will result in the two pressure gradients being the same?

5. For the time-dependent flow problem in a rigid tube of radius a, with small fluid viscosity, let the pressure gradient $G(t)$ be a rapid pulse (and zero before and after the pulse), so that the zero-order approximation for flow in the main body of the tube is

$$u_0(t) = \begin{cases} 0, & \text{for } t \leq -t_0 \text{ where } t_0 \text{ is very small;} \\ U, & \text{for } t \geq 0. \end{cases}$$

Find an approximate solution for the flow near the walls, in the form $u = u_0(t) + u_b(\xi, t)$ where $\xi = a - r$. Hence find the shearing stress on the walls. What condition must be imposed on the time t for this approximation to be valid?

6. As a model of an artery leading from the heart, consider an elastic-walled tube of resting radius r_0, wall thichness h, and elastic modulus E, attatched to a reservoir, the whole being filled with fluid of density ρ. Initially the pressure is p_0 and the fluid velocity is zero. It is required to produce a velocity

$$u(t) = \begin{cases} U \sin \omega t, & \text{for } 0 < t < \pi/\omega; \\ 0, & \text{for } t > \pi/\omega, \end{cases}$$

at the entrance to the tube ($x = 0$). Neglecting viscosity of the fluid, write down the velocity distribution along the tube at time t. Hence find the pressure variation at $x = 0$, and show that the maximim pressure is

$$p_0 + U\sqrt{E\rho h/2r_0}.$$

7. Consider a three-way junction at $x = 0$, where a large artery (radius a) suddenly becomes two identical smaller arteries (each of radius $b < a$). All other material properties are unchanged. The matching conditions across the junction are (i) continuity of pressure $p(0_-, t) = p(0_+, t)$ and (ii) continuity of volume flow

$$a^2 u(0_-, t) = 2b^2 u(0_+, t).$$

If there is no reflection at this junction, i.e., the pulse waves in all tubes travel in the $+x$ direction only, show

$$\frac{b}{a} = 2^{-2/5} = 0.76.$$

8. Show by actually solving the differential equation (8.96) given in text that the pressure $P(t)$ during the systolic phase can be written as given by (8.98).

9. Suppose the diastolic phase starts at $t = 0$ with pressure P_0 and lasts for a time t_0. Find the time t_1 of the systolic phase so that at the end of this time the pressure is again P_0.

10. If P_1 denotes the pressure at the end of systole, then show

$$P_1 = \frac{B \sin \psi + Ae^{-\alpha T_s}}{1 - e^{-\alpha T_0}},$$

where $T_0 = T_s + T_d$. If $\alpha^2 \ll \beta^2$, and approximate values of $\alpha = 1$, $T_s = 0.3\,\text{s}$, $T_d = 0.7\,\text{s}$ are assumed, show that the approximate value of P_1 is given by $2.8B$.

11. If P_0 denotes the pressure at the beginning of systole, then show

$$P_0 = \frac{B \sin \psi e^{-\alpha T_d} + Ae^{-\alpha T_0}}{1 - e^{-\alpha T_0}}.$$

12. (i) A major concern of cardiologists is the relationship between blood pressure and blood velocity in the arterial system. Under what circumstances can the velocity pulse be obtained simply as a constant multiple of the pressure pulse? Indicate (briefly) ways in which a better pressure–velocity relationship may be obtained.

(ii) A perfect pulse as in (i) is travelling in the $+x$ direction when it suddenly encounters a complete obstruction at $x = 0$, where the velocity vanishes. Show that the pressure felt at $x = 0$ is twice that of the original unreflected pulse.

13. Given that

$$\frac{\partial u}{\partial t} = -\frac{1}{\rho}\frac{\partial p}{\partial x},$$

$$\frac{\partial p}{\partial t} = -\frac{Eh}{2r_0}\frac{\partial u}{\partial x}$$

are the (linearized) momentum and mass-conservation equations for velocity and pressure of fluid (density ρ) in an elastic-walled tube (thickness h, resting radius r_0, elastic modulus E), obtain the differential equation satisfied by the pressure. What is the wave velocity, taking $E = 5 \times 10^6 \,\text{dyne cm}^{-2}$, $h/r_0 = 0.1$, $\rho = 1.0\,\text{g cm}^{-3}$?
Consider a wave travelling towards the right, in which

$$p(0, t) - p_0 = P \sin \omega t$$

for all t, where P is a constant. Find p and u for all x, t, given that the solution for u contains no steady component. Evaluate the maximum fluid velocity at any point, using the values already given, and $P = 2 \times 10^4 \, \text{dyne cm}^{-2}$.

If this wave were travelling towards the *left*, give the expression for p and u. Combine this wave with the one travelling to the right, and show that at $x = 0$, the velocity is zero, but the pressure fluctuation is

$$p(0, t) - p_0 = 2P \sin \omega t.$$

14. Suppose that the resting radius and wall thickness of an artery change at one point ($x = 0$) from r_1, h_1 to r_2, h_2, the elastic modulus E remaining the same. Assuming a pulse

$$p - p_0 = f\left(t - \frac{x}{c}\right)$$

incident at $x = 0$ from the left, use the principles of continuity of volume flow and of pressure, at $x = 0$, to find the complete solution for pressure and velocity on each side of the change. Write down the reflection and transmission coefficients. If

$$f(t) = \begin{cases} 0 & \text{for } t < 0; \\ p_1 \sin \omega t, & \text{for } 0 < t < \pi/\omega; \\ 0, & \text{for } t > \pi/\omega, \end{cases}$$

what volume of blood passes into the right-hand section?

15. The pressure inside the heart (LV) itself can be modelled by an *inhomogeneous* wave equation

$$\frac{\partial^2 p}{\partial t^2} - \frac{1}{c^2} \frac{\partial^2 p}{\partial x^2} = P(t),$$

where $P(t)$ is a given function representing pressure development by heart muscle. If $x = 0$ is the closed end where $u = 0$, and $x = L$ is the open (valve) end, where a (perfectly matched) aorta begins (i.e., $p = \rho c u$ at $x = L$), solve for p and u, $0 \le x \le L$ and in particular find the output flow pulse into the aorta, namely

$$u(L, t) = \frac{1}{2\rho c} \left\{ p(t) - P\left(t - \frac{2L}{c}\right) \right\}.$$

Note that there is a left-travelling wave as well as a right-travelling one in $0 \le x \le L$.

9

Analysis and Applications of Left Ventricular Mechanics

9.1. Analysis and Interpretation of Left Ventricular Nonlinear Constitutive Properties During Diastole and Systole

With the rapid advance in medical technology, there has been an expansive array of new research tools for the analysis and integration of physiological data. These newer techniques have facilitated physiologists to reappraise the traditional concepts underlying the evaluation and assessment of cardiac performance. For some time, function of the heart was characterized solely in terms of its pumping efficiency. However, in assessing cardiac function an understanding and knowledge of the stress developed within the wall of the left ventricle (LV) during a cardiac cycle is of paramount importance. Some of the material properties of the left ventricle, such as diastolic stiffness, are an important reflection of myocardial disease, for example, cardiomyopathy. However, variation in the regional material constitutive property can help detect the location of myocardial infarct. Left ventricular wall stress is another property which can indicate the degree of left ventricular compensation to pressure and/or volume overload, which occurs with valvular disease. For the determination of both left ventricular stress and stiffness properties, ventricular pressure and geometry data are required. At present this information can only be determined using simultaneous cardiac catheterization and cineangiocardiography for pressure and geometry, respectively.

For the purpose of mathematical analysis, the LV will be modelled as a uniformly thick-walled pressurized spherical shell of homogeneous but nonlinear material properties. The stress–strain relations for this large deformation sustaining LV model will be synthesized by using stress–strain results from uniaxial test of LV papillary muscle, whose large deformation state can be more conveniently characterized by stretch λ (which is the ratio of the length of a line segment in the deformed state to that in the undeformed state), rather than the engineering strain e (which is the ratio of the change in length of a line element due to deformation to its original length). The relationship between engineering strain and stretch, when there is no change in direction of the line segment (as in the case of uniaxial tensile test on the papillary muscle), is given by

$$e = \lambda - 1. \tag{9.1}$$

However, for deformation where the line elements are rotated, the relationship between engineering strain and stretch is more complex.

We define stress, S, here as the ratio of force to the actual deformed area on which the force acts. For the muscle (or for the myocardium), we have further to consider the condition of incompressibility of the material. If λ is the stretch along the axis of the papillary muscle of the specimen and δ along the transversal direction, then the incompressibility condition can be mathematically expressed as (Fung, 1967)

$$\lambda\delta^2 = 1. \tag{9.2}$$

If further we denote the undeformed and deformed cross-sectional areas of the muscle specimen by A and A' respectively, and if the force exerted on the specimen be denoted by F, then we have the following relationships:

$$A' = \delta^2 A = \frac{A}{\lambda} \tag{9.3}$$

and

$$S = \frac{F}{A'} = \frac{F\lambda}{A}. \tag{9.4}$$

The nonlinear stress–stretch behaviour of the papillary muscle, proposed by Valanis and Landel (1967), is given in the form

$$S_i = C\lambda_i^\alpha + \bar{p}, \tag{9.5}$$

where S_i is the ith principal stress, λ_i is the ith principal stretch, $\bar{p}$ is a hydrodynamic pressure term and C and α are material constants (or constitutive parameters). For the uniaxial papillary muscle specimen, we can therefore consider one principal direction along the axis of the specimen and other two principal directions along two perpendicular directions to the axis of the specimen. Thus, if S_1 and λ_1 denote the stress and stretch along the applied load axis, then we have from equation (9.5)

$$S_1 = C\lambda_1^{\alpha} + \bar{p}. \tag{9.6}$$

Since the other two principal stresses are zero and the two principal stretches are equal to $\lambda_1^{-1/2}$, by virtue of the incompressibility relation (9.2), we have the relationship,

$$0 = C\lambda_1^{-\alpha/2} + \bar{p}. \tag{9.7}$$

Thus, by subtracting equation (9.7) from (9.6), the uniaxial stress–stretch relation reduces to

$$S_1 = C\left(\lambda_1^{\alpha} - \lambda_1^{-\alpha/2}\right). \tag{9.8}$$

Based on the test data for rat papillary muscle by Janz and Grimm (1973), the material constants C and α have been determined to be $2.37\,\text{g cm}^{-2}$ and 18 respectively.

We will now adapt this constitutive relationship to the three-dimensional pressurized left ventricular spherical mode (Moriarty, 1980), for which the constitutive relation is given by (9.5) and the principal stretches $(\lambda_r, \lambda_\theta, \lambda_\phi)$ are given by

$$\lambda_r = \frac{dr}{dR}, \qquad \lambda_\theta = \lambda_\phi = \frac{r}{R}, \tag{9.9}$$

where R and r are the undeformed and deformed coordinates respectively. The equation of equilibrium in spherical polar coordinates is given by

$$\frac{dS_r}{dr} + \frac{2}{r}(S_r - S_\theta) = 0. \tag{9.10}$$

The incompressibility condition requires

$$\lambda_r \lambda_\theta \lambda_\phi = 1. \tag{9.11}$$

From equations (9.9) and (9.11), we get

$$\frac{r^2}{R^2}\frac{dr}{dR} = 1. \tag{9.12}$$

The solution to equation (9.12) is given by

$$r^3 = R^3 + \beta, \tag{9.13}$$

where β is a constant of integration. Thus, from equations (9.9) and (9.13), we have the following values of the principal stretches:

$$\lambda_r = \left(\frac{R^3}{R^3+\beta}\right)^{2/3}, \qquad \lambda_\theta = \lambda_\phi = \left(\frac{R^3+\beta}{R^3}\right)^{1/3}. \tag{9.14}$$

Similary, the expressions for the principal stresses, S_r, S_θ, S_ϕ can be obtained as follows. Using equation (9.5), we first obtain

$$S_r = C\lambda_r^\alpha + \bar{p} = C\left(\frac{R^3}{R^3+\beta}\right)^{2\alpha/3} + \bar{p},$$

$$S_\theta = S_\phi = C\lambda_\theta^\alpha + \bar{p} = C\left(\frac{R^3+\beta}{R^3}\right)^{\alpha/3} + \bar{p}. \tag{9.15}$$

Upon substituting these expressions into the equilibrium equation (9.10), we obtain

$$\frac{dS_r}{dr} = \frac{2C}{(R^3+\beta)^{1/3}}\left\{\left(\frac{R^3+\beta}{R^3}\right)^{\alpha/3} - \left(\frac{R^3}{R^3+\beta}\right)^{2\alpha/3}\right\}. \tag{9.16}$$

Therefore,

$$\frac{dS_r}{dR} = \frac{2CR^2}{(R^3+\beta)}\left\{\left(\frac{R^3+\beta}{R^3}\right)^{\alpha/3} - \left(\frac{R^3}{R^3+\beta}\right)^{2\alpha/3}\right\}. \tag{9.17}$$

Integrating equation (9.17) with respect to R, and using the condition that $S_r = -p$ when $R = a$ (where a is the undeformed radius of the endocardium), we obtain

$$S_r = 2CX - p, \tag{9.18}$$

where

$$X = \int_a^R \frac{R^2}{(R^3+\beta)}\left\{\left(\frac{R^3+\beta}{R^3}\right)^{\alpha/3} - \left(\frac{R^3}{R^3+\beta}\right)^{2\alpha/3}\right\} dR. \tag{9.19}$$

By satisfying the second boundary condition, $S_r = 0$ for $R = b$, in equation (9.18), we finally obtain the expression for the chamber pressure, p as follows:

$$p = 2C\int_a^b \frac{R^2}{(R^3+\beta)}\left\{\left(\frac{R^3+\beta}{R^3}\right)^{\alpha/3} - \left(\frac{R^3}{R^3+\beta}\right)^{2\alpha/3}\right\} dR, \tag{9.20}$$

in terms of the undeformed chamber radii (a, b) and the material constitutive properties (C, α). The constant of integration, β, in equation (9.13), is obtained in terms of the undeformed and deformed volumes (V_0 and V respectively) from

$$\frac{V}{V_0} = 1 + \frac{\beta}{a^3}. \tag{9.21}$$

The stress S_r is given by equation (9.18), as

$$S_r = C \int_a^R \frac{2R^2 \Phi}{(R^3 + \beta)} \, dR - p, \tag{9.18'}$$

where

$$\Phi = \left\{ \left(\frac{R^3 + \beta}{R^3} \right)^{\alpha/3} - \left(\frac{R^3}{R^3 + \beta} \right)^{2\alpha/3} \right\}. \tag{9.22}$$

The stress S_θ is then obtained from equations (9.16), (9.18) and (9.10), as

$$S_\theta = C \left\{ \int_a^R \frac{2R^2 \Phi}{(R^3 + \beta)} \, dR + \Phi \right\} - p. \tag{9.23}$$

This general theory will now be applied to determine and physiologically interpret the left ventricular constitutive properties (C and α, in equation (9.5)) and myocardial stress, for the diastolic phase, and for the early ejection phase from the instant of opening of the aortic valve to the instant of occurrence of maximum ventricular pressure.

9.2. Application to the Diastolic Phase

For the diastolic phase, the term $\bar{p}$ in the constitutive relationship (9.5) represents the myocardial tissue fluid pressure. Let (p_i, p_{i+1}), (a_i, a_{i+1}) and (b_i, b_{i+1}) be the values of the chamber pressure and the deformed inner and outer radii of the left ventricular model, at the instants i and $i + 1$ during diastole. By substituting them successively into equation (9.20), we obtain the value of α ($= \alpha_d$), from the monitored values of p_i, a_i ($= (3V_i/4\pi)^{1/3}$), b_i, p_{i+1}, a_{i+1}, and b_{i+1}:

$$\frac{\int_{a_i}^{b_i} \left(\frac{R^2}{R^3+\beta} \right) \left(\left(\frac{R^3+\beta}{R^3} \right)^{\alpha/3} - \left(\frac{R^3}{R^3+\beta} \right)^{2\alpha/3} \right) dR}{\int_{a_{i+1}}^{b_{i+1}} \left(\frac{R^2}{R^3+\beta} \right) \left(\left(\frac{R^3+\beta}{R^3} \right)^{\alpha/3} - \left(\frac{R^3}{R^3+\beta} \right)^{2\alpha/3} \right) dR} = \frac{p_i}{p_{i+1}}, \tag{9.24}$$

where β is given, with the help of equation (9.13), as

$$\beta = a_i^3 - a^3, \tag{9.25}$$

and then the value of C ($= C_d$) is obtained as

$$C = \frac{\frac{P_i}{2}}{\int_{a_{i+1}}^{b_{i+1}} \left(\frac{R^2}{R^3+\beta} \right) \left(\left(\frac{R^3+\beta}{R^3} \right)^{\alpha/3} - \left(\frac{R^3}{R^3+\beta} \right)^{2\alpha/3} \right) dR}. \tag{9.26}$$

The global values of C and α for the ventricle would be sensitive to normal and diseased states. Thus, if ranges for C and α are developed for normal and diseased myocardial left ventricles, they can be employed thereafter in differential diagnosis. Having thus determined the values of the constitutive parameters, the stresses S_r and S_θ can be evaluated by means of equations (9.18), (9.22) and (9.23).

9.3. Application to the Systolic (Early Ejection) Phase

In the systolic phase the constitutive relationship is given by equations (9.5) and (9.9), where the coordinates R and r respectively correspond to the start of ejection (when the ventricular pressure equals the aortic diastolic pressure p_d) and the instant of peak ventricular pressure (equal to the aortic systolic pressure p_s). The reason why this analysis is applied to the early ejection phase and not to the late ejection phase is that it is presumed that the contractile constitutive properties ($C = C_s$, $\alpha = \alpha_s$) may be more reasonably assumed to remain constant during the early ejection phase rather than during the late ejection phase, since the constancy of (C_s, α_s) is implicit in the analysis.

The values of the constitutive properties (C_s, α_s) of the left ventricle during systole are obtained from equations (9.24)–(9.26), where p_d and p_s are substituted for p_i and p_{i+1}, a_i and b_i correspond to the left ventricular model radii at the start of the ejection, and (a_{i+1}, b_{i+1}) are the model chamber radii at peak chamber pressure.

Now, since p_d and p_s can be determined by sphygmomanometry, the values of C_s and α_s can be obtained noninvasively. Two new indices are now proposed. The first index is C_s/p_d, which can be interpreted as the myocardial contractile stiffness required to raise the chamber pressure to p_d. The second index, α_s, may then be interpreted as an incremental stiffness parameter required to raise the chamber pressure to its peak value p_s. Together, they could be looked upon as new intrinsic indices of myocardial contractility required to sustain the arterial pressure at the value of p_s/p_d.

Again, the myocardial stresses can be evaluated by means of equations (9.22) and (9.23). In the case of stenotic aortic valves, the deviation from normality in their values can be viewed as an index of ventricular compensative hypertrophy.

Problems

1. In a two-dimensional generalized plane-stress analysis, show that

$$e_1 = \frac{1}{2}\left(\lambda_1^2 - 1\right), \qquad s_1 = \frac{E}{1 - \nu^2}\left(e_1 + \nu e_2\right),$$

$$e_2 = \frac{1}{2}\left(\lambda_2^2 - 1\right), \qquad s_2 = \frac{E}{1 - \nu^2}\left(e_2 + \nu e_1\right),$$

   where e_1 and e_2 are the principal strains, s_1 and s_2 are the principal stresses and E and ν are elastic constants.

   Show also that the ratio of the infinitesimal area element in the deformed state to that in the initial state is given by

$$\frac{dA}{dA_0} = \lambda_1 \lambda_2.$$

2. If λ_1, λ_2, λ_3 denote the principal stretches, show that the ratio of the infinitesimal volume element in the deformed state to that in the initial state is given by

$$\frac{dV}{dV_0} = \lambda_1 \lambda_2 \lambda_3.$$

   Express the volume ratio in terms of the principal strains. Show that

$$\lambda_1^2 \lambda_2^2 \lambda_3^2 = 8I_3 + 4I_2 + 2I_1 + 1,$$

   where I_1, I_2 and I_3 are the three strain invariants.

3. Assume that the left ventricle is modelled mathematically as a thick-walled hemispherical shell with constant thickness. If p_i and p_e denote the internal and external pressure exerted on the ventricle during the diastolic phase, show that the average stress on the ventricular wall is given by

$$\frac{p_i a_i^2 - p_e a_e^2}{a_e^2 - a_i^2},$$

   where a_i and a_e are respectively the inner and outer radii of the shell.

4. Show that, for a sphere with internal and external radii a_i and a_e respectively and subjected to a left ventricular pressure P, the circumferential stress at any radius r is given by

$$Pa_i^3 \frac{\left(1 + a_e^3/2r^3\right)}{\left(a_e^3 - a_i^3\right)}.$$

10

Analysis and Applications of Heart Valve Vibration

10.1. Basic Concepts

The heart valves are mechanical devices that permit the flow of blood in one direction only. The two cuspid (atrio-ventricular) valves are located between the atria and ventricles, while the two semi-lunar valves are located at the entrance to the pulmonary artery and the great aorta. The cuspid valves prevent blood from flowing back into the atria from the ventricles and the semi-lunar valves prevent it from being regurgitated back into the ventricles from the aorta and pulmonary artery.

Any one of the four valves may lose its ability to close tightly; such a condition is known as valvular insufficiency or regurgitation. On the other hand, stenosis is an abnormality in which the fully open valvular orifice becomes narrowed by scar tissue that forms as a result of valvular disease; thereby the blood flow through the valve results in increased pressure drop across the valve orifice.

Here, we will address ourselves to aspects concerning the noninvasive assessment of valvular pathology, as characterized by its nonlinear constitutive property. In order to do so, we need to incorporate the noninvasive measurements of cardiac mechanical events, associated with valvular mechanics, that are available to us. One such set of measurements constitutes monitoring the geometry of cardiac structures by means of two-dimensional echocardiography. The heart sounds constitute yet another set of noninvasive data which can be utilized.

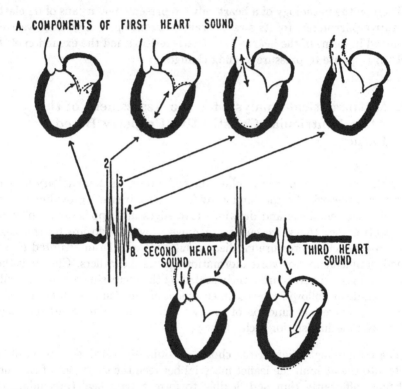

FIGURE 10.1. Schematic drawings of the causes of various components of the heart sounds

The acoustic waves (due to cardiac structual vibrations associated with cardiac mechanical events) that are transmitted to the chest contain information on the vibratory source. Each heart sound is composed of vibrations, which can be characteristically described by their intensity, frequency, duration and quality. Here, we will concern ourselves with the heart sounds caused by heart valve vibrations. The intensity and frequency of these heart sounds are respectively influenced by the rate of development of pressure differentials on either side of the valve, and by the distance of excursion of the valve leaflets along with the elasticity of the valve tissue.

By carrying out the frequency analysis of heart sounds recorded on the chest wall, by associating heart sounds experimentally and theoretically with the physics of valvular vibrations (Figure 10.1*), and by computing the range of spectral frequency of valvular model vibrations, it is possible to correlate the dynamical events of vibrating valves and the heart sound characteristics.

Thereby the pathology of a heart valve, represented by means of its elastic constitutive parameters (of its stress–strain relationship) can be noninvasively determined in terms of the heart sound characteristics and the excursion of the leaflets in response to pressure loading changes.

10.2. Mathematical Analysis for the Assessment of the Atrio-Ventricular Constitutive Property Based Pathology

During the final stages of ventricular filling, the roots of the papillary muscles are displaced towards the apex and away from the valve ring, providing tension in the chordae tendineae and drawing valve edges into apposition. Following atrial contraction, the valve cusps are moved towards closure by the negative atrio-ventricular pressure gradient resulting from the interrupted stream of blood, supplemented by vortex formation behind the leaflets. Closure is then completed, and the valves are tightly sealed by the increasing atrio-ventricular pressure gradient, following the commencement of ventricular systole. The similarity in structure and function of the two atrio-ventricular valves suggests a similar basic mechanism for each.

The restraining action of the chordae tendineae prevents eversion of the leaflets into the atrium, the leaflet material between the insertions of the chordae being sufficiently thin and flexible to form a tight seal (precluding the

---

*These drawings are based on the concept that the vibrations are induced by acceleration or deceleration of the blood within elastic chambers.

The first sound can be divided into four components. The initial vibrations occur when the first myocardial contractions in the ventricle shift towards the atrium to approximate and seal the atrio-ventricular valves. The second component begins with abrupt tension of closed atrio-ventricular valves decelerating the moving blood. It may represent oscillation of blood initiated by overdistention of the atrio-ventricular myocardium. The reaction would be similar to tapping a balloon filled with water. The third component may involve oscillations of blood between the distending root of the aorta and the ventricular walls. The fourth component probably represents vibrations due to turbulence in blood flowing rapidly through the ascending aorta and pulmonary artery. (From Rushmer, 1976.)

possibility of retrograde flow), whose static deformation is proportional to the pressure differential loading across the mitral valve. The initiation of ventricular contraction and associated disturbances in the intraventricular blood-fluid can very likely set the mitral valve into vibration and give rise to appropriate components of the first heart sound. In fact, the first major component of the first heart sound, S_1, has been observed to be coincident with mitral valve closure (Mills *et al.*, 1976).

Towards the determination of the in-vivo stiffness (modulus) of the valve leaflets, the A-V valve will be modified as an idealized membrane supported along the valve ring edge, and obeying a nonlinear stress–strain relationship. Consider the mitral valve leaflets at that stage in left ventricular systole when the valve has just closed. While the leaflets are fixed around the curve edge by the fibrous valve ring, the free edges are held in apposition to each other by the suspensory ligaments, the chordae tendineae. The Young's modulus of the chordae is significantly higher than that of the membrane tissue (Clark, 1973), so that each leaflet boundary may be regarded as being entirely fixed. Furthermore, the fluid pressure across the valves causes tension in the leaflets, which are held in a distended state and prevented from moving into the atrium by the restraining action of the chordae tendineae.

The stress (σ)–strain (ϵ) characteristics of normal mitral valve tissue can be, according to Clark (1973), expressed mathematically as

$$\sigma \text{ (dyne cm}^{-2}) = 4083.6 \left(e^{17\epsilon} - 1\right), \tag{10.1}$$

whereas the stress–strain behaviour for unhealthy and fatty mitral valve leaflet tissue can, according to Ghista and Rao (1973), be expressed in the form

$$\sigma \text{ (dyne cm}^{-2}) = 9714.6 \left(e^{17\epsilon} - 1\right). \tag{10.2}$$

Hence the leaflet constitutive can, in general, be represented by

$$\sigma = c \left(e^{17\epsilon} - 1\right), \tag{10.3}$$

where the value of the constant c characterizes the relative stiffness of the normal or diseased tissue. If we denote the elastic modulus by E ($= d\sigma/d\epsilon$), then we obtain, from equation (10.3),

$$E = 17(\sigma + c). \tag{10.4}$$

This constitutive property can be represented on the constitutive E–σ histological coordinate plane in Figure 10.2[†], where the normal and pathological domains are delineated.

[†]Based on the constitutive relations (10.4): $E = 17(\sigma + c)$, where $c \leq 4083.6$ dyne cm^{-2} for normal tissue and ≥ 4083.6 dyne cm^{-2} for pathological tissue.

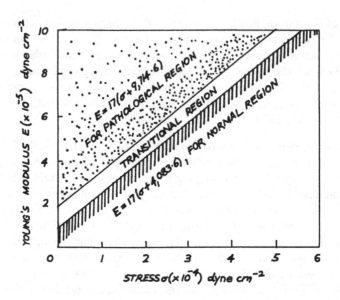

FIGURE 10.2. Division of the E–σ plane into normal and pathological regions for the mitral tissue valve

The objective of the analysis is to express the elastic modulus, or alternatively the material parameter c, in terms of the fundamental vibrational mode frequency, f_{11}, and the leaflet shape and dimensions. The detailed derivation of this relationship has been developed by Mazumdar and Hearn (1978). We thus obtain

$$c = \sigma \left(\exp \left(\frac{17 q^2 u^*}{16 T^2} \right) - 1 \right)^{-1}, \tag{10.5}$$

where q is the atrio-ventricular pressure difference, T is the membrane tension and u^* is a geometrical parameter depending upon the shape of the membrane.

In order to obtain the value of c from essentially monitorable quantities of membrane deflection and vibrational frequency, we need to substitute for σ and T in terms of these monitorable quantities. For this purpose, if we invoke the differential equation of vibration of the leaflet membrane (Mazumdar, Hearn and Ghista, 1979) we obtain the following expression for the constitutive parameter:

$$c = 13.69 f_{11}^2 u^* \left(\exp \left(\frac{4.25 W_{\max}^2}{u^*} \right) - 1 \right)^{-1}, \tag{10.6}$$

where f_{11} is the fundamental mode frequency in Hz, $W_{\max}$ is the maximum value of membrane displacement and the value of the geometric parameter u^*

is evaluated by adopting the appropriate contour function u, and tabulated in Table 10.1.

TABLE 10.1. Geometric parameter u^* for membranes of various shapes. In view of the actual size of the valve leaflet, we assume that each leaflet has an area $\pi/2$ cm^2

Membrane Model	Value of u^*
Semi-elliptic with aspect ratio	
1.0	0.195
0.9	0.184
0.8	0.170
0.7	0.155
0.6	0.137
0.5	0.117
Parabolic (aspect ratio 0.5)	0.131
Limaçon (with shape parameter $\epsilon = 1/2$)	0.203

For convenience of evaluating c from equation (10.6), nomograms have been developed for c in terms of the vibrational frequency of the membrane, for specific values of u^* and for a range of possible values of the maximal membrane displacement W_{max} (Figure 10.3). The material parameter c, evaluated from these nomograms, designates the value as normal or pathological, based on equations (10.1)–(10.4) and Figure 10.2. However, the value of c can only be obtained noninvasively from the nonograms of Figure 10.3 if the valve vibrational frequency can be justifiedly determined from the first heart sound by spectral phonocardiography.

A typical digitized phonocardiogram, shown in Figure 10.4, illustrates two high-frequency complexes in the 50–140 Hz domain. The resulting three-dimensional time-dependent *amplitude vs. frequency* spectra in Figure 10.5 (corresponding to successive instants between A and B within the first heart sound S_1 of Figure 10.4, show short durational, high-frequency peaks coincident with the first complex of high-frequency oscillation manifest in the digitized phonocardiogram (shown in Figure 10.5, referred to as the first major component of the first heart sound), and with mitral leaflet coaption.

Thus, mitral valve vibrations must be manifested in this major component of the first heart sound, particularly since its monitored frequency content in fact corresponds to the frequency range obtained from our derived expression (10.6) of mitral valve vibrational frequency (for representative values of u^*,

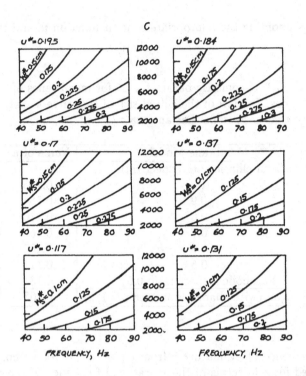

FIGURE 10.3. Values of the material parameter c, plotted as a function of fundamental vibrational frequency of the leaflet, for various values of the geometric parameter u^* and maximal static displacement W_s^* ($= W_{max}$ in the text).

W_{max} and c):

$$f_{11}^2 = \frac{c\left(\exp\left(4.25 W_{max}^2/u^*\right) - 1\right)}{13.69u^*}. \tag{10.7}$$

The distribution of the fast Fourier transform (FFT) magnitude over the spectral frequency content (Figure 10.6) shows a number of peaks, which could represent the primary and higher modes' vibrational frequencies of the left ventricular chamber and heart valve. The question now arises as to which frequency peak we should adopt as the one representing the primary vibrational mode of the mitral valve.

To this end, we rationalize that since the mitral valve constitutes one segment of the left ventricular chamber, its primary vibrational mode must be a secondary or higher vibrational mode of the left ventricle. The second vibrational mode of a left ventricular spherical model is the bending mode, which

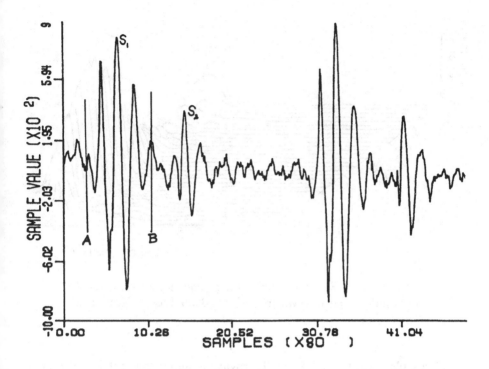

FIGURE 10.4. Digitized phonocardiogram for a subject (from Hearn, 1981)

has four nodal points. Now, since the mitral valve may be taken to constitute one-fourth of this spherical left ventricular wall, we can assume that the primary mode of vibration (and its associated vibrational frequency) of the mitral valve corresponds to the second mode of vibration of the left ventricle (and its associated vibrational frequency). Hence, the second peak frequency in the spectral plot of Figure 10.6 corresponds to the primary vibrational frequency of the mitral valve, which as seen in the plot, is 55 Hz.

Now, we can return to the clinical application of our derived expression (10.6) for the material parameter c. By firstly monitoring the valve geometry (and hence u^* from Table 10.1) and W_{max} by means of sector-scan echocardiography, and secondly adopting its primary vibrational frequency equal to that of second peak spectral frequency of the major component of the first heart sound, we can determine the value of the parameter c from the nomograms of Figure 10.3. Corresponding to this in-vivo value of c, the linear E–σ constitutive relationship for the mitral valve leaflet tissue can be designated by means of equation (10.4). The position of this line in the E–σ histological

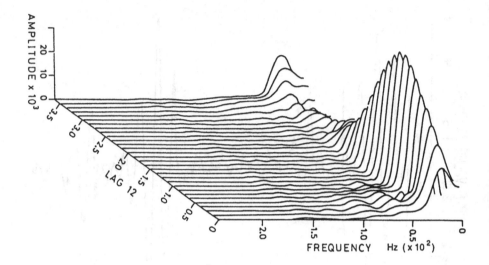

FIGURE 10.5. Successive overlaping frequency spectra, corresponding to the interval from A to B on Figure 10.4 (from Hearn, 1981)

coordinate plane (in Figure 10.2) will provide an indication of the normality or abnormality of the stiffness property and hence, of the pathology of the leaflet material.

10.3. Vibrational Analysis of Semi-lunar Valves, with Physiological Interpretation

Some investigations indicate that the second heart sound originates from vibration of the aortic valves (and pulmonary valve), the aorta (and the pulmonary artery), and the whole cardiohaemic system (Wiggers, 1915; Sabbah and Stein, 1976); this is based on the correlation of amplitude and velocity of deflection of the valve in an in-vitro system (as measured by high speed motion pictures) with the sound pressures (recorded by means of cathetertip micromanometers). Let us examine the physics of vibrations of semi-lunar valves.

Following coaption of the valve leaflets, a pressure difference is developed across the closed valve during the isovolumic relaxation phase, and the leaflets distend slightly toward the ventricle. This causes a pressure reduction within the blood in the aorta in the case of the aortic valve (or in the pulmonary artery in the case of the pulmonary valve). Subsequent recoil of the valve compresses

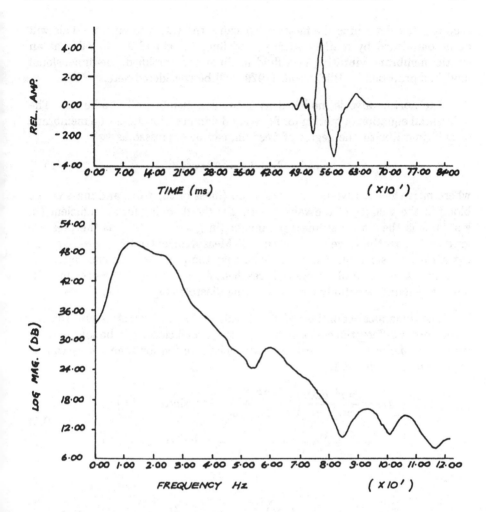

FIGURE 10.6. Phonocardiographic segment gated first heart sound and corresponding relative magnitude of FFT for a subject (from Hearn, 1981)

the blood in the aorta. Vibration of the leaflets gradually diminishes due to damping. The leaflet vibration produces transient pressure gradient in the surrounding blood medium, which subsequently causes vibration of contiguous structures, which are transmitted to the chest wall where they are recognized as audible heart sounds.

If therefore one conceives that the vibrations of the closed aortic and pulmonary valves are sufficient (if not primary) causes of the second heart sound, then one can determine the factors that cause the valves to vibrate. This will be accomplished by mathematically modelling the closed aortic valve as an elastic membrane vibrating in a fluid medium. A simplified one-dimensional model as proposed by Blick *et al.* (1979) will be considered here.

The aortic valve is modelled as a circular membrane of radius a. The differential equation of motion for forced and damped vibrations of a membrane at any time t, with one degree of freedom, can be expressed as

$$m\ddot{x} + d\dot{x} + kx = \Delta p \pi a^2, \tag{10.8}$$

where m is the effective mass of vibration (mass of the valve and mass of the blood in the vicinity of the valve, in g), d is the damping force coefficient (in g s^{-1}), k is the elastic stiffness parameter (in g s^{-2}), and Δp is the pressure gradient across the valve (in dyne cm^{-2}). Measurements of the pressure gradient across the semi-lunar valve indicate that the pressure difference increases linearly with time until a time t_1 is reached, following which the pressure difference remains essentially constant during diastole (Figure 10.7).

The dynamical equation (10.8) for valve vibration, which is an ordinary second order differential equation with constant coefficients, can be solved easily using any one of the standard methods. The solution for time $t < t_1$ can be expressed as (Stein, 1981)

$$\begin{aligned}
x(t) &= \frac{\pi a^2}{m\omega_n^2} \frac{d\Delta p}{dt}\left(t - \frac{2\xi}{\omega_n} + \frac{1}{\omega}e^{-\xi\omega_n t}\sin(\omega t - \Psi)\right) \\
&+ \frac{\pi a^2}{m\omega_n^2}\Delta p_0\left(1 + \frac{1}{(1-\xi^2)^{1/2}}e^{-\xi\omega_n t}\sin(\omega t - \Psi)\right),
\end{aligned} \tag{10.9}$$

where

$$\begin{aligned}
\omega_n &= \left(\frac{k}{m}\right)^{1/2}, \\
\xi &= \frac{d}{2m\omega_n}, \\
\omega &= \omega_n\left(1 - \xi^2\right)^{1/2}, \\
\Psi &= 2\tan^{-1}\left(\frac{(1-\xi^2)^{1/2}}{-\xi}\right),
\end{aligned} \tag{10.10}$$

and where ω_n denotes the natural frequency (in rad s^{-1}), ξ is the nondimensional damping factor, ω is the actual frequency (in rad s^{-1}), Ψ is the phase lag (in rad) and Δp_0 is the pressure difference across the valve at the moment of closure. The velocity of centreline deflection of the valve, obtained by

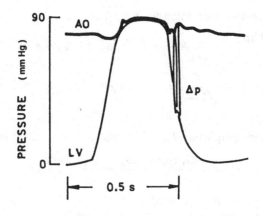

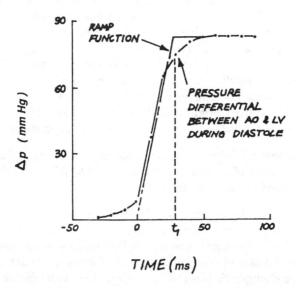

FIGURE 10.7. Top: Simultaneous recording in a dog of aortic (AO) and left ventricular (LV) pressures. Values of the pressure gradient that developed across the closed valve are indicated as Δp. Bottom: Instantaneous values of Δp are plotted with time. Prior to t_1 the curve is plotted as a straight line. Following t_1, Δp becomes nearly constant. The ramp function that approximates the curves is shown. Reproduced with the kind permission of P. Stein

differentiating the above equation for $t < t_1$, is

$$\dot{x}(t) = \frac{\pi a^2}{m\omega_n^2} \frac{d\Delta p}{dt} \left(1 - \frac{\xi}{\sqrt{1-\xi^2}} e^{-\xi\omega_n t} \sin(\omega t - \Psi) + e^{-\xi\omega_n t} \cos(\omega t - \Psi) \right)$$
$$+ \frac{\pi a^2}{m\omega_n^2} \Delta p_0 \left(-\frac{\xi}{\sqrt{1-\xi^2}} e^{-\xi\omega_n t} \sin(\omega t - \Psi) + e^{-\xi\omega_n t} \cos(\omega t - \Psi) \right).$$

$$(10.11)$$

The sound pressure amplitude P_{amp} is given (in dyne cm^{-2}) by,

$$P_{amp} = \frac{G_1 \pi \rho \omega a^4}{RK} \frac{d\Delta p}{dt} + \frac{G_2 \pi \rho \omega a^4}{R\sqrt{km}} \Delta p, \qquad (10.12)$$

where G_1 and G_2 are proportionality constants, R is the distance from the valve to the site of measurement and ρ is the density of blood (in g cm^{-3}).

The solution of the modelling equations (10.8) is determined for representative values of Δp (assumed to be constant at 150 mmHg s^{-1} for time equal to or greater than 0.0175 s), valve stiffness ($k = 8.8 \times 10^6$ dyne cm^{-1}), damping force factor ($d = 2.8 \times 10^3$ dyne s cm^{-2}), effective mass ($m = 195$g) and the rate of pressure difference Δp. The solutions for $x(t)$ and $\dot{x}(t)$ show good correlations with the in-vitro experimental measurements of the vibrations of a normal stent-mounted porcine valve incorporated in a hydraulic chamber in the cardiovascular system (Figure 10.8). It can be seen that the theoretical response curves are similar to the curve derived from experimental data.

The above analysis helps to explain many previously unexplained clinical observations by means of factors that are identified to relate to valve vibration and the heart sounds that result from it.

Identification of the influence of the stiffness factor, k, in valve vibration and sound production can explain (by dint of the term ω_n^2 in equation (10.9)) the diminished second sound in calcified aortic stenosis, and its continued presence in the case of congenital stenosis of aortic valve not yet damaged by degenerative changes (Sabbah et al., 1978). This same factor explains an increased amplitude of the pulmonary component of the second sound relative to the aortic component in pulmonary hypertension (Stein et al., 1978).

Another important factor is the effect of the viscosity factor, d, on the valve vibration (equations (10.9) and (10.11)). Since anaemic patients have reduced viscosities, they are shown to have augmented heart sounds (Stein et al., 1978).

Also, from the relation $\omega_n = \sqrt{k/m}$, it is clear that the higher the stiffness of the valve leaflet, the higher will be the natural frequency of vibration. Hence, patients with aortic stenosis can be expected to have a higher frequency of the

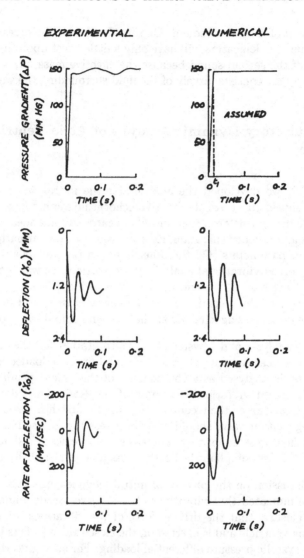

FIGURE 10.8. The pressure gradient Δp (top), centreline deflection x_0 (centre), and rate of centreline deflection $\dot{x}_0$ (bottom) as they varied with time after valve closure. On the left are actual values measured in-vitro. On the right are values calculated from the equations descriptive of x_0 and $\dot{x}_0$, assuming that Δp varied as the ramp function shown at the upper top. Reproduced with permission of P. Stein

aortic component of the second sound. On the other hand, the increased mass of valve leaflet due to calcification will have only a little effect upon the frequency or amplitude of the second sound because the effective mass, m, appearing in the expression of ω_n consists largely of the fluid surrounding the valve leaflets.

10.4. Auscultatory Dynamic Analysis of Atrio-Ventricular Valves

In Section 10.2, we developed the analysis for the mitral valve, which was specifically designed to utilize the in-vitro echocardiographic measurement of valvular geometry and deformation and phonocardiographic measurements of first heart sound frequency spectrum, for evaluation of valvular pathology from the constitutive parameters. On the other hand, in the previous section, Section 10.3, our forced vibrational analysis of the aortic valve was modelled as a spring-dashpot mechanical system, in order to provide qualitative physiological correlations with the second heart sound characteristics and bring to bear insights on the factors influencing second heart sound amplitude and frequency.

A similar analysis and physiological interpretation for the mitral valve leaflet will now be carried out, wherein the initial pressure-loaded mitral valve leaflet vibration is analysed as a consequence of the time-dependent pressure loading. This forced vibrational response of the leaflets will determine the nature and magnitude of their contribution to the first heart sound S_1. For instance, a regurgitant valve should contribute less towards sound production due to diminished pressure loading, whereas gross calcification should produce less sound as the less compliant leaflets are less able to elastically recoil.

In our discussion on the physics of mitral valve leaflet vibration (in Section 10.2), we indicated that vibrations of A-V valves occur immediately following valve closure, that the driving force of the vibrations originates from the contracting ventricle and is exerted on the closed valve leaflets by means of rapidly rising systolic pressure differential loading. For an engineering mechanics representation of this process, the atrio-ventricular leaflet is modelled immediately following closure as a deformable, thin, elastic membrane immersed in a viscous fluid medium, subject to a time-dependent pressure loading, and having an initial velocity which corresponds to the rate of closure of the leaflet. The response of the membrane to this pressure loading will determine the extent and nature of vibrational response; the associated sound pressure amplitude will then be expressed in terms of this response.

The general two-dimensional equation of motion for forced and damped vibrations of a membrane at any time τ can be expressed as (Mazumdar, 1973),

$$T \oint \frac{\partial W}{\partial n} \, ds - d \iint \frac{\partial W}{\partial \tau} \, d\Omega + \iint F \, d\Omega = \rho \iint \frac{\partial^2 W}{\partial \tau^2} \, d\Omega, \qquad (10.13)$$

where W is the membrane displacement, being a suitable function of the iso-amplitude contours $u(x, y) = $ constant. In the above equation, the various terms appearing are due to elastic forces, viscous damping, driving forces and inertial forces, respectively; the coefficients T, d and ρ are respectively the membrane tension, the coefficient of viscous damping and the density.

The solution to equation (10.13) can be obtained using the normal mode expansion in terms of the eigenfunctions of the associated free vibration problem. Thus, if W_i and ω_i denote the eigenfunctions and the eigenfrequencies of the corresponding free vibration problem, then the solution, W, can be expressed as a linear sum of the eigenfunctions W_i in the form

$$W = \sum_{i=1}^{\infty} g_i(\tau) W_i(u). \qquad (10.14)$$

If the driving force function F, appearing in equation (10.13), is considered to be time-dependent only then we have

$$F(x, y, \tau) = q h(\tau), \qquad (10.15)$$

where q is a unit pressure loading. Then, using a normal mode expansion in terms of eigenfunctions, it is possible to express the driving force function in the form

$$F = h(\tau) \sum_{i=1}^{\infty} a_i W_i(u), \qquad (10.16)$$

where the coefficients a_i can be obtained using the orthogonality relations of the eigenfunctions. The values of the coefficients a_i are thus obtained as

$$a_i = \frac{2}{B_i} J_1(B_i), \qquad (10.17)$$

where B_i is the ith zero of the zeroth order Bessel function J_0, and J_1 is the first order Bessel function. Using the orthogonality condition, the differential equation for $g_i(\tau)$ is obtained in the form (Hearn and Mazumdar, 1981)

$$\ddot{g}_i(\tau) + \frac{d}{\rho} \dot{g}_i(\tau) + \omega_i^2 g_i(\tau) = \frac{a_i}{\rho} h(\tau). \qquad (10.18)$$

Clearly, the solution of the above equation will depend on the form of the forcing function $h(\tau)$.

The mitral valve sustains a rapidly increasing pressure loading (which is the difference in pressure between the left atrium and left ventricle), following atrio-ventricular pressure equalization at the commencement of ventricular systole. This pressure difference increases with time until it attains a peak of approximately 100 mmHg (Figure 10.9a). Experimental studies of Laniado *et al.* (1973) indicate that mitral valve closure occurs some 40 ms after the atrio-ventricular pressure crossover (from positive to negative) so that the mitral leaflets will be subjected to an initial nonzero pressure loading at closure of approximately 20 mmHg.

Accordingly, the forcing function $h(\tau)$, in equation (10.15), is reasonably approximated by the ramp function (Figure 10.9b), whereby the mitral valve is subjected to an initial pressure loading of magnitude S, followed by a linearly rising pressure gradient $dp/d\tau$ up to time τ_1 and thereafter by a constant pressure. Thus, the approximated form for the input pressure loading is

$$h(\tau) = SH(\tau) + \frac{dp}{d\tau} \int_0^\tau \left(H(t) - H(t - \tau_1) \right) dt, \qquad (10.19)$$

where $H(\tau)$ is the Heaviside unit step function.

It is clear from the governing dynamic analysis that the magnitude of the time derivative of the pressure loading will have a significant influence on the velocity of the valve leaflet motion; any factor which alters the rate of development of the pressure gradient across the leaflet may then be expected to affect the heart sound intensity accordingly. Such factors may include ventricular contractility as well as the timing of closure with respect to ventricular systole. In support of this, there exists experimental evidence of the correlation of the amplitude of the first group of high-frequency vibration of S_1 with left ventricular $dp/d\tau$ during isovolumic systole (Sakamoto *et al.*, 1965). Contribution to S_1 intensity from the tricuspid valve response to right ventricular systole is also expected (Mills *et al.*, 1976).

Figures 10.10a and 10.10b illustrate left- and right-sided atrio-ventricular valve responses; therein, the tricuspid valve response has been calculated for a systolic pressure loading of 25% of the left ventricle, for the following representative data: mitral leaflet fundamental natural frequency of 50 Hz; damping-to-mass ratio of 14.36 dynes cm^{-1} g (following Blick *et al.*, 1979); the input parameters S, $dp/d\tau$ and τ_1 as 20 mmHg, 5 mmHg ms^{-1}, and 20 ms, respectively; systolic pressure loading of 25% of that in the left ventricle.

For computing the tricuspid leaflets' response, the relative lower blood mass of the right ventricle and relatively smaller size of the tricuspid leaflets prompted the choice of higher natural frequency (65 Hz), higher damping-to-mass ratio, and an initial pressure loading of 5 mmHg reaching a peak of

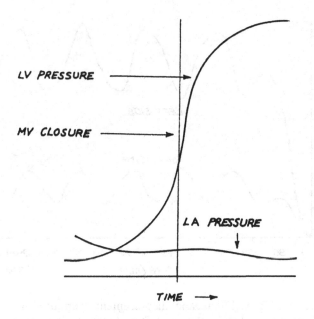

FIGURE 10.9a. Left ventricular and atrial pressure curves

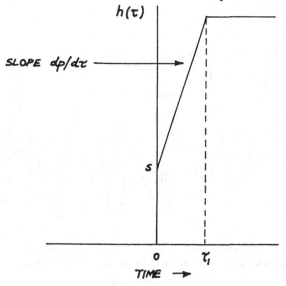

FIGURE 10.9b. The ramp approximation (equation (10.19)) of the systolic pressure loading sustained by the mitral valve

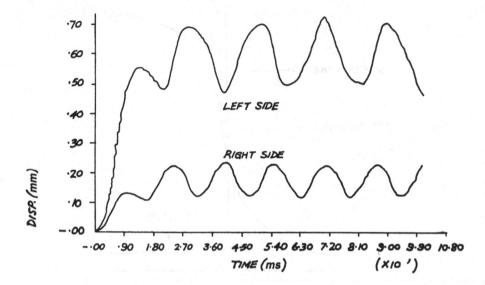

FIGURE 10.10a. Membrane displacement computed at the point of maximal deflection for left and right ventricular pressure loadings

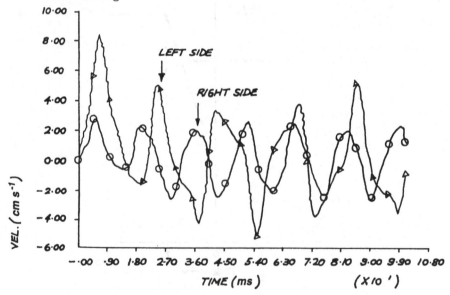

FIGURE 10.10b. Velocity computed at the point of maximal deflection for left and right ventricular pressure loadings

25 mmHg in 0.02 s. It is noted that the appreciably reduced displacement and velocity responses of the tricuspid valve results in a much lower rate of energy dissipation compared to its left ventricular counterpart, thereby confirming that the first heart sound is predominantly of left-sided origin (Luisada et al., 1967).

For further work in this area, it would be interesting to compute and plot, firstly the left ventricular contractility index $dp/d\tau$ vs. the membrane displacement velocity and the vibrational sound pressure amplitude for different values of membrane tension and density, and secondly the relation of the vibrational response vs. blood viscosity.

10.5. Comments on Clinical Correlations

If vibration of the mitral valve is the cause of the mitral component of the first heart sound, then the driving force productive of vibration would clearly be the pressure difference across the valve (Stein, 1981). In the case of a circular, tensed membrane, the extent and velocity of deflection and amplitude of the sound produced from it would be dependent upon the rate at which the driving force develops. If a pressure gradient exists across the valve at the moment of valve closure, this pressure drop would be part of the driving force. The rate of change of the systolic pressure gradient across the closed mitral valve, therefore, can be approximated by the rate of rise of left ventricular pressure (dp/dt). It has been shown that maximal left ventricular dp/dt relates closely to the amplitude of the first sound (Sakamoto et al., 1965), and this effect can be explained in terms of the driving force productive of valve vibration.

Another clinical manifestation that can be explained on the basis of the rate of exchange of the pressure gradient across the valve is the accentuated first sound in patients with noncalcific mitral stenosis. If the valve is distensible, the accentuated first sound in mitral stenosis may relate to the rate of change of the pressure gradient across the mitral valve at the moment of mitral valve closure. Due to the elevated left atrial pressure in mitral stenosis, the mitral valve closes when the rate of change of left ventricular pressure (dp/dt) is more rapid than it would be with a normal valve and normal left atrial pressure (Sabbah et al., 1978). If the mitral valve closes when left ventricular dp/dt is relatively high, then a more rapidly developing driving pressure across the mitral valve would occur and produce a more forceful vibration. For this reason, the mitral component of the first heart sound could be accentuated. Both the amplitude of the first sound and left ventricular dp/dt at the time of onset of the mitral component of the first heart sound have been shown to be higher in patients with mitral stenosis than in normal subjects (Stein, 1981). This suggests that the rate of change of pressure gradient across the closed mitral valve at the moment of closure was higher in patients with mitral stenosis. Perhaps, the pressure difference between the left ventricle and left atrium was also higher at the moment of valve closure and this may have contributed to the accentuated first heart sound.

The first heart sound is accentuated in patients with large PR interval, and this may relate to the magnitude of dp/dt at the time of valve closure and the time of initiation of systolic vibrations of the valve. If the mitral valve has already floated to a closed or nearly closed position with prolongation of the PR interval, then the initial forces that cause distention of the valve at the time of complete closure would be relatively small and the first sound would be accentuated.

If the mitral valve is incapable of vibration as it would be if it were rigid, then the first sound would be absent or markedly diminished in intensity. This is a well-recognized entity in patients with calcific mitral stenosis. A comparable effect was clearly present in patients with aortic stenosis, and the mechanism was shown to be due to diminished or absent valve vibration (Sabbah and Stein, 1976; Sabbah *et al.*, 1978).

The first sound is shown to be accentuated in patients with a low blood viscosity (anaemic patients). This suggests that the viscosity of blood may have a damping effect upon vibrations produced by the mitral valve, just as it has with the semi-lunar valves. Also, animals with mitral valves of larger than normal cross-sectional area are said to have higher intensity first sound (Stein, 1981). This is compatible with the larger surface of an acoustical source producing a more intense sound.

In conclusion, there is a variety of evidence which, when integrated, suggests that the first sound is due to vibration of the closed mitral valve and tricuspid valves. Other vibrations, if they occur, are secondary. Preliminary modelling of vibration of the closed mitral leaflets identifies the factors shown to affect the amplitude of the first sound, and many auscultatory observations can be interpreted on the basis of the factors identified in this mathematical model of valve vibration.

Problems

1. Show that the constitutive parameter c can be expressed as

$$c = 13.69 f_{11}^2 u^* \left(\exp\left\{ \frac{0.0057 q^2}{\rho^2 f_{11}^4 u^*} \right\} - 1 \right)^{-1},$$

   where u^* is the maximum value of iso-amplitude contour lines $u(x,y) = $ const, ρ is the density per unit area of the valve leaflet, q is the atrio-ventricular pressure difference and f_{11} is the fundamental mode frequency in Hz.

2. Derive the expression for the stress in the valve leaflet as a function of the pressure loading q across the valve leaflet in the form

$$\sigma = \left(\frac{E q^2 u^*}{16 h^2 (1 - \nu)} \right)^{1/3},$$

   where ν is the Poisson ratio of the leaflet material.

3. Deduce the relationship between the leaflet modulus E and the frequency f_{11} in the form

$$E = \frac{32 \pi^2 D (1 - \nu) f_{11}^2 u^{*2}}{(2.4049)^2 W_{\max}^2},$$

   where it is assumed $q = -2 T W_{\max} / u^*$.

4. Solve the differential equation (10.8) for the aortic valve vibration if

$$\Delta P = P(1 - e^{-rt})$$

   where P is the maximum pressure difference between the aorta and the left ventricle. Compare the solution thus obtained with the solution obtained in the text where ΔP is considered as a ramp function.

5. Solve the differential equation (10.18) for mitral valve vibration where pressure loading is given by equation (10.19).

6. Discuss the above problem when the pressure loading is expressed in the form of an exponential function as in Problem 4.

11

Medical Devices

"Medical device" is a recent terminology used to identify a wide variety of medical equipments. Medical devices are generally classified into three principal categories: diagnostic, therapeutic, and assistive (or rehabilitative).

11.1. Diagnostic Devices

Diagnostic devices are very common because of their usefulness when a physician first examines a patient. Perhaps the two most familiar diagnostic devices are the thermometer and the stethoscope. A thermometer measures body temperature, which is an indicator of the process of regulated burning of body fuel called metabolism. The stethoscope collects the feeble sounds made by some internal body organs and presents these sounds to the physician's ear. The timing and nature of these sounds give important clues to the physician regarding the function of these organs. A stethoscope is also used to measure blood pressure when an occluding cuff is wrapped around the upper arm and quickly inflated to a pressure high enough to occlude the artery under it. When the pressure in the cuff is lowered slowly, blood spurts through the collapsing artery, and a stethoscope is used to monitor the downstream arterial sounds, the appearance and disappearance of which indicate when to read the cuff pressure to determine the maximum (systolic) and minimum (diastolic) blood pressure.

There are some organs in the body which function in relatively inaccessible locations. These organs must be transilluminated for visualization. The low-energy x-ray beam is used for this purpose, but because of the insensitiveness of the human eye to electromagnetic energy, a photographic surface or fluorescent screen is used to convert the absorption image to a visible one. Sometimes, low-energy ultrasound waves are also used to produce images of internal structures. Similarly, photographs of radiant heat emitted by the body, known as thermograms, indicate the qualitative nature of the temperature distribution on the various parts of the body surface.

There are many organs, viz. heart, brain, muscles, nerves, eyes, ears, etc., which produce signals due to electric potentials associated with them. These low-amplitude signals can be enlarged and displayed on paper tape or on oscilloscope screens. The results, known as ECG, EMG, EEG, etc., yield important diagnostic information.

11.2. Therapeutic Devices

After the diagnosis is made, then comes the therapeutic stage. Drugs are the most extensively used therapy, but even with this form of therapy, instruments are of importance because many drugs are introduced into the body with the aid of instruments, the most commonly used one being the needle and syringe.

Therapeutic devices are used to control or cure an illness. A number of familiar and unfamiliar therapeutic devices are in use these days. The most familiar one is the short-wavelength x-ray beam which kills the growth of cancerous cells in the body. The low-wavelength infrared spectrum is used for heating tissues within the body, which enhances circulation in the irradiated area. The high-intensity coherent beam of the laser is now used as a precise cautery to spot-weld detached retinas and to perform surgical cuttings. Another commonly used therapeutic device is the electrosurgical instrument permitting surgeons to cut and coagulate tissue which is difficult to excise due to excessive bleeding.

Ultrasonic instruments which operate in the megahertz to a fraction of a megahertz range are used with low intensity to produce heat and alter metabolism and circulation at the cellular level. With high intensity, a cutting action is produced, and the effect has been exploited in surgery.

A variety of devices that produce an electrical output are used therapeutically. For example, the electrical stimulator, which generates pulses of current, is used to exercise paralyzed muscles while nerve regeneration is taking place. In order to arrest ventricular fibrillation when there is a cessation of the pumping action of the heart and loss of blood pressure, a current pulse is passed through the chest wall. Also, for certain psychiatric disorders, an electric current pulse is passed through the head as shock therapy. When the ventricles of the heart lose their drive and start beating slowly, single shocks delivered rhythmically to the ventricles then produce contractions. Such stimulators are called cardiac pacemakers. Most pacemakers are permanently implanted in the heart.

Another therapeutic device is the "artificial kidney" known as the dialyzer. Its main function is to remove or add substances to the bloodstream while the diseased kidneys are recovering. At present, home dialysis is becoming an increasingly popular technique.

Another implanted therapeutic device that has been developed recently is what is called a "cochlea implant" or the "bionic ear". The cochlea is one of the most astounding organs in the body. It consists of a long coiled tube filled with fluid. Sound waves from outside, relayed and amplified by the eardrum and a chain of three tiny bones, set up ripples in the fluid and a host of tiny hairs along the walls of the cochlea are set moving back and forth. The movement of each hair actuates impulses in a nerve fibre, corresponding to a certain frequency or pitch of sound. The surgery for this implant is extensive. Bone from the skull is cut away to form a cavity of 2.5 cm diameter and about 5 cm deep. Into that is placed a silicon rubber mould holding a radio-receiver. From there, a bundle of 22 exceedingly fine wires or electrodes is threaded through the inner ear or cochlea to be linked directly to some of the fibres of the auditory nerve which normally carries impulses to the brain. The implanted receiver is only half the bionic ear. The rest is worn on the body with a microphone behind one ear, a programmable speech processor carried in a pocket, and a transmitter held against the head by a band to send signals to the receiver. The speech processor is the key of the device and it is in this area that real scientific progress is likely to take place by mathematicians and computing specialists.

11.3. Assistive or Rehabilitative Devices

After an illness has been cured, most often assistive devices are needed to compensate for a deficit in normal functioning of the system. Furthermore, with the rapid progress of medical technology, the percentage of aged people increases. As a result assistive devices for this group become of paramount importance. Perhaps the most familiar assistive device is the hearing aid. Next comes the assistive device for the blind person. There is a great need for such devices. Several guidance devices for blind persons, such as an optical-distance measuring instrument, a portable sonar, and a capacity-type proximity indicator, are now on the way.

Another category of assistive devices quite familiar to people are crutches and walkers that are used when control of the muscles of locomotion is impaired. Artificial arms or legs are other types of assistive devices that are now becoming very beneficial to many handicapped people.

Those who are interested in medical progress – mathematicians, computing scientists, physicists and biomedical engineers – can find challenging problems for themselves in all these areas of medical endeavour.

Glossary of Terms

AIDS – acquired immunodeficiency syndrome.

Aorta – the main trunk from which the systemic arteries originate.

Atrium – an upper chamber on either side of the heart.

Auscultation – listening for sounds produced within the body.

Axon – the long outgrowth of the body of a nerve cell.

Carriers – individuals who harbour infection which can be transmitted to others.

Catheter – a tubular instrument passed through body channels for withdrawl or introduction of body fluids.

Cineangiography – the photographic recording of fluoroscopic images of the organ by motion picture techniques.

Diastole – the phase of the cardiac cycle in which the heart relaxes between contractions.

DNA – deoxyribonucleic acid found in chromosomes.

Echocardiography – recordings of internal structures of the heart by the echo obtained from ultrasound beams.

Electrocardiography – the graphic recording of the electric currents generated by the heart. Tracing is called ECG.

Electroencephalography – the recording of changes in electric potentials of the brain. Tracing is called EEG.

Electromyography – the recording of the intrinsic electrical properties of muscle. Tracing is called EMG.

Erythrocyte – a red blood cell, or corpuscle.

Flowmeter – an apparatus for measuring the rate of flow of liquids or gases.

HIV – human immunodeficiency virus.

Hypertension – persistently high blood pressure.

In vitro – within a glass.

In vivo – within the living body.

Leukocyte – a colourless blood corpuscle.

Lumen – the cavity or channel within a tubular organ, as a blood vessel or the intestine.

Metabolism – the process by which in any living body, human or plants, food is converted into energy.

Myocardium – the thickest layer of the heart wall.

Noninvasive – without any surgical intervention.

Pharmacokinetics – the study of the action of drugs on living systems.

Phonocardiography – the graphic recording of heart sounds.

RNA – ribonucleic acid found in chromosomes.

Sphygmomanometer – an instrument for measuring arterial blood pressure.

Stenosis – narrowing of a blood passage or opening.

Systole – the period of the cardiac cycle in which the heart contracts.

Thrombosis – formation or presence of blood clots inside a blood vessel.

Ultrasound – mechanical energy of a frequency greater than 20 000 cycles per second.

Vasoconstrictor – causing constriction of the blood vessels.

Vasodilator – causing dilation of blood vessels.

Ventricle – a small cavity or chamber, as in the brain or heart.

References

Abé, H., Nakamura, T., Motomiya, M., Konno, K. and Arai, S. (1978). Stresses in left ventricular wall and biaxial stress–strain relation of the cardiac muscle fiber for potassium-arrested heart. *J. Biomech. Eng., Trans. ASME* **100**: 116–121.

Anderson, R.M. (1988). The epidemiology of HIV infection: variable incubation plus infectious periods and heterogeneity in sexual behaviour. *J. Roy. Statist. Soc.* **A 151**: 66–93.

Anderson, R.M. and May, R.M. (1989). Complex dynamical behaviour in the interaction between HIV and the immune system. In *Cell to Cell Signaling* (Goldbeter, A., ed.). Academic Press, New York: 335–349.

Anderson, R.M., Meddley, G.F., May, R.M. and Johnson, A.M. (1986). A preliminary study of the transmission dynamics of the human immunodeficiency virus (HIV), the causative agent of AIDS. *J. Maths. Appl. in Medicine and Biol.* **3**: 229–263.

Anliker, M., Rockwell, R.L. and Ogden, E. (1971). Nonlinear analysis of flow pulses and shock waves in arteries. *Z. Angew. Math. Physics* **22**: 217–246, 563–581.

Attinger, E.O. (1964). *Pulsatile Blood Flow*. McGraw–Hill, New York.

Baker, D.W. and Daigle, R.E. (1977). Noninvasive ultrasonic flowmetry. In *Cardiovascular Flow Dynamics and Measurements* (N.H.C. Hwang and N.A. Norman, eds.). University Park Press, Baltimore.

Bartlett, M.S. and Hiorns, R.W. (1973). *The Mathematical Theory of the Dynamics of Biological Populations*. Academic Press, London.

Benaceraff, B. and Unanue, E.R. (1979). *Textbook of Immunology.* University Park Press, Baltimore.

Bergel, D.H. (1972). *Cardiovascular Fluid Dynamics.* Academic Press, Vols. 1 and 2.

Blake, J.R. (1975). On the movement of mucus in the lung. *J. Biomech.* **8**: 179–190.

Blake, J.R. (1977). On the fluid mechanics of the foetal lung. *Math. Scientist* **2**: 95–110.

Blick, E.F., Sabbah, H.N. and Stein, P.D. (1979). One-dimensional model of diastolic semilunar vibrations procedure of heart sounds. *J. Biomech.* **12**: 223–227.

Blower, S.M., Hartel, D., Dowlabatai, H. and Anderson, R.M. (1991). Drugs, Sex and HIV: A Mathematical Model for New York City. *Phil. Trans. Roy. Soc. London Ser. B, Biol. Sci.* **331**, 1260: 171–187.

Brown, R. (1828). *Philosophical Magazine* **4**: 161.

Caro, C.G., Pedley, T.J., Schroter, R.C. and Seed, W.A. (1978). *The Mechanics of the Circulation.* Oxford University Press, Oxford.

Carslaw, H.S. and Jaeger, J.C. (1959). *Conduction of Heat in Solids.* 2nd edition, Oxford University Press, Oxford.

Chavez, C. (1989). *On the role of long incubation periods in the dynamics of acquired immunodeficiency syndrome.* Lecture Notes in Biomathematics **81**, Springer, New York.

Clark, R.E. (1973). Stress–strain characteristics of fresh and frozen human aortic and mitral leaflets and chordae tendineae. *J. Thoracic and Cardiovascular Surgery* **66**: 202–208.

Cohen, J.E. (1968). *A Model of Simple Competition.* Harvard Univeristy Press, Cambridge.

Cox, C.B., Healey, I.N. and Moore, P.D. (1976). *Biogeography. An Ecological and Evolutionary Approach.* 2nd edition, Blackwell Scientific Publications, Oxford.

Crank, J. (1956). *The Mathematics of Diffusion.* Oxford University Press, Oxford.

Delisi, C. (1976). *Antigen–Antibody Interactions.* Lecture Notes in Biomathematics 8, Springer, Heidelberg.

Diamond, J.M. (1975). The island dilemma: Lessons of modern biogeographical studies for the design of natural reserves. *Biol. Conserv.* 7: 129–146.

Dinnar, U. (1981). *Cardiovascular Fluid Dynamics.* CRC Press, Baton Rouge, FL.

Einstein, A. (1956). Investigations on the theory of brownian movement (English translation by A.D. Cowper). In *A. Einstein* (R. Fürth, ed.). Dover, New York.

Fick, A. (1855). The Law of Diffusion. *Ann. Phys. Leipzig.* 170: 59.

Folkow, B. and Neil, E. (1971). *Circulation.* Oxford University Press, New York.

Forrester, J.H. and Young, D.F. (1970). Flow through a converging-diverging tube and its implications in occlusive vascular disease. *J. Biomech.* 3: 297–316.

Frank, O. (1899). Die Grundform des arteriellen Pulses. *Z. Biol.* 37: 483–526.

Fung, Y.C. (1967). Elasticity of soft tissues in simple elongation. *Am. J. Physiol.* 213: 1532–1534.

Fung, Y.C. (1981). *Biomechanics, Mechanical Properties of Living Tissues.* Springer-Verlag, New York.

Fung, Y.C. (1984). *Biodynamics, Circulation.* Springer-Verlag, New York.

Ghista, D.N. (1979). *Applied Physiological Mechanics.* Harwood Academic Publishers, New York.

Ghista, D.N. and Rao, A.P. (1973). Mitral valve mechanics – Stress–strain characteristics of excised leaflets. Analysis of its functional mechanics and its medical application. *Med. Biol. Eng.* **11**: 691–702.

Gilpin, M.E. (1975). Limit cycles in competition communities. *Am. Nat.* **109**: 51–60.

Gilpin, M.E. and Diamond, J.M (1976). Calculation of immigration and extinction curves from the species-area-distance relation. *Proc. Natl. Acad. Sc. U.S.* **73**: 4130–4134.

Goodman, T.R. (1958). The heat-balance integral and its application to problems involving a change of phase. *ASME Trans.* **80**: 335–342.

Hahn, B.H. and Shaw, G.M. (1986). Genetic variation in HTLV - III/LAV over time in patients with AIDS or at risk of AIDS. *Science* **232**: 1548–1533.

Hearn, T.C. (1980). *A Mathematical Study of Atrio-Ventricular Valve Vibrations.* Ph.D. Thesis, University of Adelaide, Australia.

Hearn, T.C. and Mazumdar, J. (1981). A study of the dynamic response of atrio-ventricular valves using a membrane model. *Int. J. Math. Modelling* **2**: 97–107.

Hethcote, H. and Van Ark (1992). *Modelling HIV transmission and AIDS in the USA.* Lecture Notes in Biomathematics **95**, Springer.

Higgs, A.J. (1981). Island biogeography theory and nature reserve design. *J. Biogeography* **8**: 117–124.

Hwang, N.H.C. and Norman, N.A. (1977). *Cardiovascular Flow Dynamics and Measurements.* Univeristy Park Press, Baltimore.

Ingram, D. and Bloch, R.F. (1986). *Mathematical Methods in Medicine - 1 (Statistical and Analytical Techniques).* John Wiley, New York.

Jacquez, J.A. (1972). *Compartmental Analysis in Biology and Medicine.* Elsevier, New York.

Janz, R.F. and Grimm, A.F. (1973). Deformation of the diastolic left ventricle. *Biophys. J.* **13**: 689–709.

Jerne, N. (1973). The immune system, *Scientific American* **229**: 51–60.

Jones, R.T. (1969). Blood flow. In *Annual Review of Fluid Mechanics* (W.R. Sears and M. Van Dyke, eds.). Annual Reviews, Palo Alto, California.

Kawai, C. (1981). Reconstruction of 3-D images of pulsating left ventricle from 2-D sector scan echocardiograms of apical long axis view. *Computers in Cardiology*, IEEE Computer Society, Long Beach, California.

Keyfitz, N. (1968). *Introduction to the Mathematics of Population.* Addison–Wesley Publishing Co., Reading, Mass.

Kirschner, D. (1996). Using mathematics to understand HIV immune dynamics. *Notices, American Mathematical Society* **43**: 191–202.

Lack, D. (1973). The numbers and species of hummingbirds in the West Indies. *Am. Stat.* **27**: 326–337.

Laniado, S., Yellin, E.L., Miller, H. and Frater, R.W.M. (1973). Temporal relation of the first heart sound to closure of the mitral valve. *Circulation* **47**: 1006–1014.

Lee, C.S.F. and Talbot, L. (1979). A fluid mechanical study on the closure of heart valves. *J. Fluid Mech.* **91**: 41–63.

Lighthill, J. (1975). *Mathematical Biofluid Dynamics.* SIAM Publications, Philadelphia.

Lighthill, J. (1978). *Waves in Fluids.* Cambridge University Press, Cambridge, UK.

Lin, C.C. and Segel, L.A. (1974). *Mathematics Applied to Deterministic Problems in the Natural Sciences.* Macmillan, New York.

Lotka, A.J. (1924). *Elements of Physical Biology.* Williams and Wilkins, Baltimore.

Luisada, A.A., Kurz, H., Slodki, S.J., MacCanon, D.M. and Krol, B. (1967). Normal first heart sounds with non-functional tricuspid valve or right ventricle: Clinical and experimental evidence. *Circulation* **35**: 119–125.

MacArthur, R.H. and Wilson, E.O. (1967). *The Theory of Island Biogeography.* Princeton Univeristy Press, Princeton.

McDonald, D.A. (1974). *Blood Flow in Arteries.* Williams and Wilkins, Baltimore, USA.

Malthus, T.R. (1798). *An Essay on the Principles of Populations.* St. Paul's, London.

Marchuk, G.I. (1993). *Mathematical Modelling in Immunology.* Optimization Software Inc., New York.

May, R.M. (1975). Island biogeography and the design of wildlife preserves. *Nature* **254**: 117.

May, R.M. (1975). *Stability and Complexity in Model Ecosystems.* Princeton University Press, Princeton.

May, R.M. (1981). *Theoretical Ecology.* Blackwell Scientific Publications, Saunders, Philadelphia.

Mazumdar, J. (1992). *Biofluid Mechanics.* World Scientific, Singapore.

Mazumdar, J. (1973). Transverse vibration of membranes of arbitrary shape by the method of constant deflection contours. *J. Sound. Vib.* **27**: 47–57.

Mazumdar, J. and Hearn, T.C. (1978). Mathematical analysis of mitral valve leaflets. *J. Biomech.* **11**: 291–296.

Mazumdar, J., Hearn, T.C. and Ghista, D.N. (1979). Determination of in vivo constitutive properties and normal–pathogenic states of mitral valve leaflets and L.V. myocardium. In *Applied Physiological Mechanics* (Ghista, D.N., ed.). Harwood Academic Publishers, New York.

Mazumdar, J. and Thalassoudis, K. (1983). A mathematical model for the study of flow through disc-type prosthetic heart valves. *Medical and Biological Eng. and Comput.* **21**: 400–411.

Mazumdar, J. and Woodard-Knight, D. (1984). A mathematical study of semilunar valve vibration. *J. Biomech.* **17**: 639–641.

Merchant, G.J. and Mazumdar, J. (1986). A numerical model for non-Newtonian blood flow, *Automedica*, **7**: 159–177.

Mills, P.G., Chamusco, R.F., Moos, S. and Craige, E. (1976). Echophonocardiographic studies of the contribution of the atrio-ventricular valves to the first heart sound. *Circulation* **54**: 944–951.

Mirsky, I. (1973). Ventricular and arterial wall stresses based on large deformation analyses. *Biophys. J.* **13**: 1141.

Mirsky, I. (1976). Assessment of passive elastic stiffness of cardiac muscle, mathematical concepts, physiologic and clinical considerations, directions of future research. *Prog. Cardiovasc. Dis.* **18**: 277–308.

Mirsky, I., Ghista, D.N. and Sandler, H. (1974). *Cardiac Mechanics: Physiological, Clinical and Mathematical Considerations*. John Wiley & Sons, New York.

Morgan, B.E. and Young, D.F. (1974). An integral method for the analysis of flow in arterial stenoses. *Bulletin of Mathematical Biology* **36**: 39–53.

Moriarty, T.F. (1980). The Law of Laplace, its limitations as a relation for diastolic pressure, volume, or wall stress of the left ventricle. *Circ. Res.* **46**: 321–331.

Murray, J.D. (1989) *Mathematical Biology.* Biomathematics Texts, Springer-Verlag.

Noodergraaf, A. (1978). *Circulatory System Dynamics.* Academic Press, New York.

Nossal, G.J.V. (1969). *Antibodies and Immunity.* Basic Books, New York.

Nowak, M.A. and McLean, A.R. (1991). A mathematical model of vaccination against HIV to prevent the development of AIDS. *Proc. Roy. Soc. London, Ser. B, Biol. Sci.* **246**, 1316: 141–146.

Nowak, M.A. and May, R.M. (1993). AIDS pathogenesis: Mathematical models of HIV and SIV infections. *AIDS* **7**, Suppl. 1: 513–518.

Nozyce, M., Hoberman, M., Arpadi, S., Wiznia, A. Lambert, G., Dobroszycki, J., Chang, C.L. and Louis, St.Y. (1994). A 12-month study of the effects of oral AZT on neurodevelopmental functioning in a cohort of vertically HIV-infected inner-city children. *AIDS* **8**: 635–639.

Patel, D.J. and Vaishnav, R.N. (1980). *Basic Hemodynamics and its Role in Disease Process*. Univeristy Park Press, Baltimore.

Pedley, T.J. (1980). *The Fluid Mechanics of Large Blood Vessels*. Cambridge University Press, London.

Peterman T.A., Drotman D.P., Curran J.A. (1981). Epidemiology of the acquired immunodeficiency syndrome (AIDS). *Epidemiology Rev.* **7**: 7–21.

Pielow, E.C. (1969). *An Introduction to Mathematical Ecology*. Wiley–Interscience, New York.

Rashevsky, N. (1964). *Some Medical Aspects of Mathematical Biology*. Thomas, C.C. Springfield, Ill., USA.

Remington, J.W., Noback, C.R., Hamilton, W.F. and Gold, J.J. (1948). Volume elasticity characteristics of the human aorta and the prediction of the stroke volume from the pressure pulse. *Am. J. Physiol.* **153**: 298–308.

Riggs, D.S. (1963). *The Mathematical Approach to Physiological Problems*. Williams and Wilkins, Baltimore.

Roston, S. (1959). Mathematical formulation of cardiovascular dynamics by use of the Laplace transform. *Bull. Math. Biophys.* **21**: 1–11.

Roston, S. (1962). Blood pressure and the cardiovascular system. *Ann. N.Y. Acad. Sci.* **96**.

Roston, S. and Leight, L. (1959). A practical study of the air chamber model of the cardiovascular system. *J. Clin. Invest.* **38**: 777–783.

Rubinow, S.I. and Keller, J.B. (1972). Flow of a viscous fluid through an elastic tube with application to blood flow. *J. Theor. Biology* **35**: 299–313.

Rushmer, R.F. (1976). *Cardiovascular Dynamics.* 3rd Edition, Saunders, Philadelphia.

Sabbah, H.N., Khaja, F., Anbe, D.T., Folger, G.M. and Stein, P.D. (1978). Determination of the amplitude of the aortic component of the second heart sound in aortic stenosis. *Am. J. Cardiol.* **41**: 830–835.

Sabbah, H.N. and Stein, P.D. (1976). Investigation of the theory and mechanism of the origin of the second heart sound. *Circ. Res.* **39**: 874–882.

Sakamoto, T., Kusukawa, R., MacCanon, D.M. and Luisada, A.A. (1965). Hemodynamic determinants of the amplitude of the first heart sound. *Circulation* **16**: 45–57.

Schulzer, M., Radhamani, M.P., Grzybowski, S., Mak, E. and Fitzgerald, J.M. (1994). A mathematical model for the prediction of the impact of HIV infection on tuberculosis. *Int. J. Epidemiology* **23**, 2: 400–407.

Sikarskie, D.L., Girolamo, P. and Stein, P.D. (1979). A mathematical model of aortic valve vibration. *Proc. 23rd Ann. Conf. Eng. Med. Biol.* **231**: 59.

Stein, P.D. (1981). *A Physical and Physiological Basis for the Interpretation of Cardiac Auscultation.* Futura Publishing Co., New York.

Stein, P.D. and Sabbah, H.N. (1978). Accentuation of heart sounds in anemia: An effect of blood viscosity. *Am. J. Physiol.: Heart Circ. Physiol.* **4**: 664–669.

Terborgh, J. W. (1974) Faunal equilibria and the design of wildlife preserves. In F. Golley and E. Medina (eds), *Tropical Ecological Systems: Trends in Terrestrial and Aquatic Research.* Springer, New York.

Thalassoudis, K. and Mazumdar, J. (1984) Mathematical model for turbulent flow through a disc-type prosthetic heart valve. *Medical and Biological Eng. and Comput.* **22**: 529–536.

Valanis, K.S. and Landel, R.F. (1967). The strain energy function of a hypere-lastic material in terms of the extension ratios. *J. App. Phys.* **38**: 2997–3002.

Volterra, V. (1926). *Memoria della R. Accademia Nazionale dei Lincei.* Trans-lation: Animal Ecology. McGraw–Hill, New York.

Waltman, P. and Butz (1977). A threshhold model of antigen–antibody dy-namics. *J. Theor. Biol.* **65**: 499–512.

Weyer, J. and Eggers, H.J. (1992). On the structure of the epidemic spread of AIDS: The influence of an infectious coagent. *Int. J. Med. Microbiol.* **272**, 1: 52–67.

Wiggers, C.J. (1915). *Circulation in Health and Disease.* Lee and Feliger, Phil-adelphia.

Womersley, J.R. (1957). *The Mathematical Analysis of the Arterial Circulation in a State of Oscillatory Motion.* U.S. Air Force W.A.D.C. Report TR 56–614.

Yoganathan, A.P., Gupta, R. and Corcoran, W.H. (1976). Fast Fourier trans-form in the analysis of biomedical data. *Med. Biol. Eng.* **14**: 239–244.

Young, D.F. (1968). Effect of a time-dependent stenosis on flow through a tube. *J. Eng. Ind., Trans. ASME* **90**: 248–254.

Index